Water and Plants

TERTIARY LEVEL BIOLOGY

A series covering selected areas of biology at advanced undergraduate level. While designed specifically for course options at this level within Universities and Polytechnics, the series will be of great value to specialists and research workers in other fields who require a knowledge of the essentials of a subject.

Other titles in the series:

Experimentation in Biology	Ridgman
Methods in Experimental Biology	Ralph
Visceral Muscle	Huddart and Hunt
Biological Membranes	Harrison and Lunt
Comparative Immunobiology	Manning and Turner

TERTIARY BIOLOGY SERIES

Water and Plants

HANS MEIDNER, Ph.D.

Professor of Biology
University of Stirling

and

DAVID W. SHERIFF, Ph.D.

Research Fellow
Department of Environmental Biology
The Australian National University
Canberra

Blackie
Glasgow and London

Blackie & Son Limited
Bishopbriggs
Glasgow G64 2NZ

450/452 Edgware Road
London W2 1EG

International Standard Book Numbers

Paperback 0 216 90080 8
Hardback 0 216 90081 6

Printed in Great Britain by
Thomson Litho Ltd., East Kilbride, Scotland

Preface

DURING THE LAST FIFTEEN YEARS, NEW APPROACHES TO PLANT-WATER relations have become accepted in the English-speaking world. Most of the books dealing with these new approaches are at an advanced level, and we feel that the time has come to write a book on plant-water relations for undergraduate students. As the title suggests, we choose to look at the topic from the outside, i.e. with the remarkable properties of water foremost in our minds. From this basis, we proceed to observe how the plant has evolved to fit itself into this water system. The order of topics is therefore somewhat unusual. This is open to criticism: for instance, it might be thought more logical to deal first with cell-water relations, because these form the basis for whole plant water relations. We cannot argue against this, but have nevertheless reversed the order, dealing first with water and water vapour in relation to plants, and proceeding thereafter "from the top down". Thus we begin by writing about the atmosphere and the leaves exposed to it. We then go down via petioles and stems to the roots in the soil, and conclude with the water relations of individual cells and tissues. By adopting this sequence, we do not wish to belittle osmotic aspects of plant-water relations but rather to emphasize the purely physical aspects arising from the properties of water and water vapour which have such a profound influence on plant life.

Space restrictions imposed some limitations on the treatment of the subject matter. We think that we have covered most of the relevant topics and apportioned adequate space to different sections. Chapter 1 may be thought to deal with the physico-chemical aspects of water in too detailed a manner but, since we feel that these form the very basis of plant-water relations, we have dealt with them somewhat extensively. Chapter 3 analyses in detail many of the less familiar concepts of plant-water relations and devotes considerable space to the water *reservoirs* in plants—a topic often neglected and yet of profound importance to an understanding of phenomena more recently discovered.

We are conscious of a small amount of repetition in the text as well as of many cross references; we think that in combination both of these make the text more readable without loss of precision. Some emphasis has been laid on quantitative treatments of plant-water relations and on experimental work. We wish that space had permitted the inclusion of more relevant anatomical features which should never be separated from physiology. Although we have taken the latest papers into account in an endeavour to be up to date, references do not include research papers and are restricted to advanced texts.

HANS MEIDNER
DAVID W. SHERIFF
Department of Biology
University of Stirling

Contents

CHAPTER ONE

PROPERTIES OF WATER

SOME OF THE PHYSICAL AND CHEMICAL PROPERTIES OF WATER ARE immediately relevant to plant life. A brief discussion of these properties is therefore necessary in order to state the basis of our approach to the topic of "water and plants".

1.1. Molecular properties

Most of those properties of water which we discuss below are indicative of the existence of comparatively strong attachment forces between

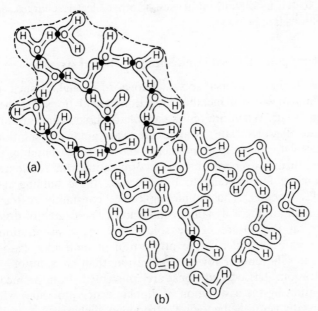

(a)

(b)

Figure 1.1. Diagrammatic representation of hydrogen-bonded clusters of water molecules (*a*) and of non-associated separate water molecules (*b*); only one hydrogen bond happens to be present.

individual water molecules. Indeed, the fact that water exists in the liquid state between $0\,°C$ and $100\,°C$ confirms this view, because other comparable molecular structures such as the hydrides of the elements close to oxygen in the periodic table (e.g. H_2S, HF and CH_4) exist at these temperatures in the gaseous state. The strongest intermolecular forces in water are hydrogen bonds. These exist when imperfectly-screened hydrogen atoms in water molecules are attached to other atoms: in the case of water molecules, they are attached to a lone pair of electrons of oxygen atoms. The particular geometry of the water molecule also confers on it an electrical dipole moment, and thereby a tendency for each molecule to associate with others via its hydrogen bonding potential.

The existence of molecular associations held together by hydrogen bonds gives liquid water a *structure*, although this structure is not fixed and the hydrogen bonds probably change continuously. This structural property of water is the basis of *cohesion forces* and the *tensile strength* of water columns in xylem vessels to which we shall refer in chapter 3 (see Experiment E1.1). However, the great electrical reactivity of the water molecule, i.e. its dipole moment, makes the structure of water subject to the modifying influences of other electric charges, especially those of ions (see below).

1.2. Solvent properties and involvement in colloidal systems

Water is the best universal solvent known to us and, as such, provides the medium in which all metabolic reactions and all transport phenomena in plants occur. When molecules capable of dissociation are in contact with water, they dissociate to some degree when going into solution. Water molecules with their dipole moments and hydrogen bonding potential interpose themselves between the ions originating from solute molecules, with bonding forces equal to or exceeding those holding the solute molecules together. Thus solute molecules constantly dissociate and reform, maintaining a dynamic equilibrium. The degree of dissociation depends on the nature of the solute and the concentration of the solution; as a general rule the proportion of molecules dissociated is greater in solutions of low concentration than in solutions of high concentration. Dissociated ions are prevented from immediate recombination by the association of dipolar water molecules with them, thus partially neutralizing their electropotential charges.

Looked at in another way, it is the structure of bulk water that is affected by the presence of solute ions. It could be said that cations such as K^+ tend to polarize the structure of water and, in the cases of Mg^{2+} and Al^{3+} for instance, four or six water molecules respectively will form "hydrated" ions; likewise, hydration occurs with anions, leading to the formation of hydroxonium ions when an H^+ of an acid is associated with a water molecule. Amphoteric colloidal particles also will cause disturbances in the structure of bulk water, when they adsorb water molecules and become electrically charged. This adsorption is especially pronounced on proteinaceous colloidal particles such as enzymes, and results in regular arrangements of water molecules (figure 1.1) in the liquid state, resembling the crystalline pattern of the solid state, ice.

Both in hydrated ions and in adsorption layers around colloidal particles, water exists not in its free association of molecules but in a more definite "bound" structural state approaching its crystalline form in the solid state. This is of the greatest importance to plant life because, not only is water the medium in which all metabolic reactions occur, but its very structure contributes to the mechanism of such reactions. In their adsorbed state, water molecules are more densely packed, i.e. at the molecular level where enzyme-catalyzed reactions occur, the molecular concentration of water is increased, and the random thermal movement of its molecules is restricted as they are held in specific orientation. This affects bond strengths within molecules and thus reactivity (Expt. E1.2; E1.3).

1.3. Surface tension and capillarity

A specific orientation of molecules occurs also at surfaces of water and interfaces with water. Here, their hydrogen bonds are located in such a way that they point inwards towards the bulk of water, and it is this that confers on liquid water the remarkably high surface tension of $73.5 \times 10^{-3} \, kg \, s^{-2}$ at $15\,°C$ (see Expt. E1.4). Surface tension is a force per unit length; but since force = mass × acceleration with dimensions $kg \times m \times s^{-2}$, surface tension has the dimensions of $kg \times s^{-2}$. (The unit of force is the newton, with dimensions $kg \, m \, s^{-2}$; thus force $\times m^{-1} = kg \, m \, s^{-2} \, m^{-1} = kg \, s^{-2}$ or Nm^{-1}.)

Looked at from the point of view of the energy relations of the system, molecules at the surface must possess an excess of energy over their fellows in the bulk of the system in order to remain at the surface against the tendency to be pulled inwards. Thus we speak of *surface energy*, and since the internal energy (p. 7) of a system tends at all times to decrease (Second Law of Thermodynamics) surfaces tend to shrink, i.e. to occupy a minimum area. This can be observed readily in the shape of dew drops (Expt. E1.4).

As a consequence of its high surface tension combined with strong adhesion forces, water has a tendency to fill capillary spaces of even comparatively large diameter. Such spaces exist between and within soil particles, cell walls, membranes, cell lumens and intercellular spaces where the presence of water is essential for one metabolic function or another (chapter 6). Water columns in capillary spaces can occur in opposition to gravity. Such capillary rise will proceed so long as the adhesion forces between water and the molecules of the wall material forming the capillary tube are stronger than the cohesion forces within the water. Whilst capillarity should not be held directly or solely responsible for water movement in plants, it does contribute to the complex mechanism that can account for the supply of water to all parts of a plant. These forces are also responsible for the retention of water droplets on leaves during mist, dew formation and rain, and following guttation (chapter 5).

The high surface tension of water and its powerful bonding with other substances have a bearing on its vaporization: they slow it down and thereby contribute towards the existence of comparatively moist habitats. In spite of this circumstance, we shall see (chapters 2, 3) that the dangers of plant desiccation due to water vapour loss are very real; indeed, most of this book deals with the complex phenomena which have evolved and which ensure supply and retention of water in plants.

1.4. Latent and specific heat

In section 1.3 we referred to the vaporization of water; for plant life this process is of the most fundamental importance. Vaporization can be considered as one of the consequences of the kinetic properties of matter. It represents a change in phase—another is fusion or, in the case of water, ice formation. For water, the energy required to bring about these changes in phase reflects once more the "structural" properties discussed above. The change in phase from solid to liquid water involves the expenditure of $333.6 \, J \, g^{-1}$. This input of energy does not bring with it a change in temperature but only the change in phase, and it is a potent sign that the "structural" change from ice to liquid water in itself requires this energy input (see Expt. E1.5). Even more spectacular is the energy requirement of $2500 \, J \, g^{-1}$ at $0 °C$, and of $2441 \, J \, g^{-1}$ at $25 °C$, to bring about the change from liquid water to water vapour. From this fact we can conclude that the internal energy (p. 7) of water vapour depends not only on its sensible heat content (temperature), but also on a considerable "latent heat" content.

Both the latent heat of fusion and, more especially, the latent heat of vaporization are of the greatest consequence for plant life. Vaporization represents one of the major mechanisms preventing the overheating of leaves as they absorb radiant energy. This is true in spite of statements to the contrary to be found in many textbooks. For every gram of water transpired at $25 °C$, the leaf tissue and its surroundings loose $2441 \, J$ of the energy gained from the sun (see Expt. E1.6), the radiation load being $80,000 \, J \, m^{-2} \, min^{-1}$. Other heat losses due to convection and turbulent air movements are also very important.

A closely related and similarly exceptional property of liquid water is its specific heat of $4·2 \, J \, g^{-1}$, almost constant between $0 °C$ and $50 °C$. For plant shoots containing usually more than 90% of water, this high specific heat prevents the occurrence of sudden and pronounced changes in their temperature during exposure to the continuous radiation so essential for photosynthesis.

1.5. Viscosity, density and gas solubilities in liquid water

(a) *Viscosity*. Viscosity is a measure of the difficulty with which molecules "slide" over one another. It is not necessarily a function of the magnitude of intermolecular forces. Indeed, water with its strong intermolecular forces has a relatively low viscosity; this is due to the hydrogen bonds which allow the molecules to slide over each other, whilst at the same time preventing their being pulled apart. Since temperature affects intermolecular forces, we would expect the viscosity of water with its high specific heat to be influenced by changes in temperature. At $5 °C$ the viscosity of water is $1·5$ centipoise (poise $= kg \, m^{-1} \, s^{-1} \, 10^{-1}$), reducing gradually to less than half the value, $0·719$ centipoise, at $35 °C$. In considering water translocation phenomena in plants and soils, these changes may be of some importance; however, as has been said above, in absolute terms the viscosity of water is low.

(b) *Density*. The density changes of water with changes in temperature are a well-known anomaly, with the greatest density at $4 °C$ in the liquid

6 PROPERTIES OF WATER

phase. Although of vital significance in biology in general, there are no special implications for plant life that need concern us here.

(c) *Solubilities*. It has been mentioned earlier (p. 4) that the all-pervading presence of water, often due to its high surface tension, adhesive properties and consequent capillary functions, is vital for plant life. One aspect of this point of view is connected with the solubilities of gases in water, especially carbon dioxide and oxygen. For carbon dioxide to reach the chloroplasts it must dissolve in water, and changes in its solubility with changes in temperature are therefore of interest and may on occasion enter calculations of "photosynthetic efficiency". An adequate oxygen supply to respiring tissues is equally important, and one portion of its diffusion path must again be in a watery medium; the solubility of oxygen in water is comparatively low, as will be seen from Table 1.1.

Table 1.1. Gas solubilities in water at different temperatures

10 °C	20 °C	30 °C	
0·0514	0·0365	0·0267	mole l^{-1} carbon dioxide
0·00017	0·00013	0·00010	mole l^{-1} oxygen

Diffusion coefficient of CO_2 in water at 20 °C $\quad 1·7 \times 10^{-9} m^2 s^{-1}$
Diffusion coefficient of O_2 in water at 25 °C $\quad 2·9 \times 10^{-9} m^2 s^{-1}$

1.6. Properties of water vapour

Properties discussed so far have been chiefly those of liquid water, except for the latent heat of vaporization, which marks the change in phase from liquid to vapour and involves fundamentals of energy relations in systems.

Thermodynamic considerations must be introduced when in the next chapter we discuss movement of water vapour from plants into the atmosphere. Therefore we begin by looking at evaporation and the vapour phase with these energy concepts in mind.

The unit of *energy* is the *joule* (J); its dimensions are $kg m^2 s^{-2}$
The unit of *force* is the *newton* (N); its dimensions are $kg m\ s^{-2}$
Work is defined as *force × distance*; its dimensions are $J = N m$

All change is the result of forces acting on a system and, in so doing, performing work. Work performed on water results in water movement. The capacity to do work is called *energy*; hence we speak of the energy content of a system. Energy exists in several interconvertible forms, e.g. chemical, electrical, kinetic, potential and thermal; of these, thermal energy or heat is an exceptional form because it cannot be converted completely into other forms of energy, while it is possible to convert other forms completely into heat. When any energy transformation occurs, the energy content of the system as a whole remains constant (First Law of Thermodynamics). However, only part of this energy content is capable of being transformed from one form to another and is thus available to do work, i.e. produce movement. This part is the internal energy of the system, and it always decreases when a change occurs spontaneously. Thus when a system undergoes change, its energy content is affected in a particular manner which we shall discuss presently, but we must first examine two possible kinds of change.

Spontaneous Irreversible Processes. It is possible to postulate a change in a system occurring in a series of infinitesimal and imperceptibly small steps, so that the system undergoing this kind of change remains in equilibrium with its surroundings, i.e. the change occurs at constant temperature. In practice, such a process cannot occur and would be known as a *reversible process*, because it would proceed without any energy transformations and could proceed in either direction with equal chance.

Changes that do occur spontaneously in nature are *irreversible*. The spontaneous evaporation of water is such an irreversible process, because the expenditure of the latent heat of vaporization required for the process must produce changes in other parts of the system—namely, in the sensible heat content of either the water itself, or its surroundings, or both. Once a state of equilibrium has been attained, the system will not spontaneously depart from this state (Second Law of Thermodynamics: *all natural or spontaneous processes, i.e. processes occurring without external interference, are irreversible in character.*)

Entropy. Having thus defined the conditions of spontaneous irreversible processes or changes in a system, we can now look more closely at the evaporation of water from the point of view of energy relations. The word *entropy* comes from the Greek word for *change*, and is thermodynamically defined as the *amount of heat absorbed by a system undergoing change, divided by the absolute temperature*. Its units are thus $J K^{-1}$, and its dimensions $kg\,m^2\,s^{-2}\,K^{-1}$; measures of entropy refer to 1 mole of a substance.

Spontaneous irreversible processes are always accompanied by an increase in entropy, an increase in molecular disorder, and a decrease in the state of molecular organization or ordered arrangement. The change from liquid water to water vapour represents a particularly pronounced decrease in orderly arrangement; thus water vapour has a particularly pronounced entropy or latent heat content but, compared with liquid water, its internal energy is reduced by several orders of magnitude.

Saturation Water Vapour Density. Considered from the point of view of the kinetic theory of matter, evaporation occurs because of the tendency for pure liquid water to establish a dynamic equilibrium with the water vapour concentration in the atmosphere in contact with it. At standard pressure and in a closed system, the equilibrium water vapour concentration over pure water will be at a specific partial pressure or the so-called *saturation water vapour pressure* (p. 40). Table 1.2 shows that the saturation water vapour pressure increases with increasing temperature. At the critical temperature (in the case of water, 100 °C), the vapour pressure of liquid water is the same as the saturation water vapour pressure of the atmosphere. The critical temperature for water at a pressure of 1 bar is 100 °C and, above the critical temperature for a given pressure, liquid water cannot exist.

Table 1.2. Saturation water vapour pressures (SWVP) and the corresponding saturation water vapour densities (SWVD) at different temperatures

	5 °C	10 °C	15 °C	20 °C	25 °C	30 °C	35 °C
SWVP (mbar)	8·72	12·27	17·04	23·37	31·67	42·43	56·23
SWVD (g m^{-3})	6·74	9·39	12·83	17·30	23·00	30·38	39·63

For all practical purposes, atmospheric pressure changes at any particular place have no effect on partial water vapour densities. On the other hand, changes in pressure in the liquid phase have a considerable effect on the partial water vapour density above the liquid. This can be observed readily in capillary systems, to which we have referred above. Here the water vapour density above the capillary meniscus must be less than that above the liquid level from which the capillary rise occurred. This difference in water vapour density must be balanced by the upward force of surface tension acting on the cross-sectional capillary circumference. This relationship is commonly used in calculations of either surface tension forces or heights of capillary rise (pp. 28, 91).

Boundary Layer and Diffusive Resistance. In an open system, the saturation water vapour density may be established virtually immediately above the water surface, in the so-called *boundary layer* (pp. 21, 34); but a small distance away from the surface the water vapour density will be less than the saturation density, and hence a water vapour density gradient will exist in the direction away from the water surface.

It is important in this context to realize that water vapour movement occurs because of the existence of density differences between two points. The forces acting to bring about this work, which results in movement of vapour down that gradient, originate in the difference in internal energy of the vapour, for instance at saturation density and at a lesser density. The resulting vapour movements are diffusive fluxes. Their rates are directly proportional to the vapour pressure differences between two points and inversely proportional to the resistance offered by the path through which the vapour has to move. If the path were a narrow tube, obviously the rate of vapour movement would be slower than if it were a wide one. We shall say more about the resistance of diffusion paths later (pp. 30–34). Resistances offered by diffusion paths assume great importance since, other evaporative conditions being equal, they alone determine the rate of diffusion (p. 33) of water vapour away from water surfaces and out of plants.

Each gas or vapour has a characteristic *diffusion coefficient* or *diffusivity*: it is slightly affected by temperature, and that for water vapour at standard pressure changes from $2.41 \times 10^{-5} \, m^2 \, s^{-1}$ at $10\,°C$ to $2.57 \times 10^{-5} \, m^2 \, s^{-1}$ at $20\,°C$ and to $2.73 \times 10^{-5} \, m^2 \, s^{-1}$ at $30\,°C$. The diffusion coefficient is the mass of a gas or vapour diffusing in unit time through a path of unit cross-sectional area and unit length, under unit difference in density.

(*Note*. The diffusion coefficients for carbon dioxide and oxygen *in water* quoted in Table 1.1 are four orders of magnitude smaller than those for the gaseous phase of water vapour shown above.)

1.7. Water potential

It will be desirable to quantify and specify the different kinds of forces acting on water in the soil, in the plant, and in the atmosphere, and to express these forces in the same units wherever possible or convenient. This can be done by defining a new term: *water potential*. By considering water potentials we can specify the direction and motion of the water. Water potential is derived from the chemical potential of water, and for those readers equipped to follow the calculations involved in this derivation we outline them in an Appendix (p. 136).

For our immediate purpose, we merely need to know that the chemical potential of pure free water at STP (μ_w^o) is at its maximum value but is not measurable in absolute terms. For this reason it is conveniently set equal to zero. The chemical potential of water (μ_w) under a (hydrostatic) tension, in an osmotic solution, adsorbed on to colloidal matter (matric) or at a lower temperature will be less than that of pure free water at STP. The difference between the chemical potential of pure free water and that under any of the conditions mentioned above ($\mu_w - \mu_w^o$) is an indication of the ability of water under any of these conditions to do work, compared with that of pure free water. For this reason, water potential has been defined either as

$$\psi = \frac{\mu_w - \mu_w^o}{\overline{V}_w} \text{ bar (force per unit area)} \tag{1.1}$$

$$\text{Water potential} = \frac{\left(\begin{array}{c}\text{chemical potential of}\\\text{water in the system}\end{array}\right) - \left(\begin{array}{c}\text{chemical potential of}\\\text{pure water at STP}\end{array}\right)}{\text{partial molar volume of water in the system}}$$

or as

$$\psi = \frac{\mu_w - \mu_w^o}{V_w^o} \text{ bar (force per unit area)} \tag{1.2}$$

$$\text{Water potential} = \frac{\left(\begin{array}{c}\text{chemical potential of}\\\text{water in the system}\end{array}\right) - \left(\begin{array}{c}\text{chemical potential of}\\\text{pure water at STP}\end{array}\right)}{\text{partial molar volume of pure water}}$$

Numerically, there is relatively little difference between calculations using equations 1.1 or 1.2. Theoretically equation 1.1 is more accurate, as it is derived directly from the chemical potential. However, equation 1.2 is normally used in practice, because the partial molar volume of pure water is accurately known; whereas the partial molar volumes of different phases* involved in a system are measurable only with extreme difficulty. The use of V_w^o in equation 1.2 complicates the relationship between water potential and osmotic pressure (Π). For an ideal solution the relationship is

---

* A thermodynamic phase is any homogeneous and physically distinct part of a system which is separated from other parts of the system by definite bounding surfaces. Thus ice, liquid water, and water vapour are three phases, and two immiscible liquids constitute two phases.

$$\psi = -\frac{\overline{V}_w}{V_w^o} \Pi \text{ bar} \qquad (1.3)$$

$$\text{Water} \atop \text{potential} = \frac{\left(\begin{array}{c}\text{partial molar volume of}\\ \text{water in system}\end{array}\right)}{\left(\begin{array}{c}\text{partial molar volume of}\\ \text{pure water}\end{array}\right)} \times \left(\begin{array}{c}\text{osmotic pressure}\\ \text{of solution}\end{array}\right)$$

If pure water and a solution are separated by a differentially permeable membrane, there will be a net flux of water molecules across the membrane into the solution. This is the process of osmosis. A pressure can be applied to the solution to prevent the net water flux (to counterbalance the osmotic forces). This is the osmotic pressure. From the ideal gas law and the derivation of chemical potential (see Appendix), it can be shown that

$$\Pi = -\frac{RT}{\overline{V}_w}\ln N_w = -\frac{RT}{\overline{V}_w}\ln a_w = -\psi \qquad (1.4)$$

where Π is the osmotic pressure, R is the gas constant ($8\cdot31\,\text{J K}^{-1}\,\text{mol}^{-1}$), T is the absolute temperature (K), $\overline{V}_w$ is the partial molar volume of water in the system, N_w is the mole fraction of water, a_w is the activity* of water, and ψ is the water potential.

In plant and soil water systems, the water potential is derived not only from the osmotic potential (ψ_π) but also from the pressure (ψ_p) and matric potentials (ψ_m). These are, either singly or added together, numerically equal to the pressure that must be applied to the system to prevent the movement of water molecules through a membrane from pure water into the system. (N.B. Osmotic and matric potentials operate in the opposite direction to the counter-balancing pressures, and are thus always negative. Pressure potentials, when negative, represent tensions; when positive, hydrostatic pressures.)

One of the main uses of water potentials is to ascertain the *direction* and volume of water flow between any two points. The flow of water in the soil-plant-air system has been considered as analogous to the flow of an electric current. This means that Ohm's law can be applied to determine the volume of water moving between two points.

---

* *Activity* is the property of a real solution that takes the place of the mole fraction of an ideal solution in the free energy equation; this is necessary because all concentrated solutions deviate from the ideal. The activity can be defined as: *a hypothetical solution of unit molar concentration possessing the properties of a very dilute solution.*

$$Q = \frac{V}{R} \qquad (1.5)$$

$$\left(\begin{array}{c}\text{volume of water flowing}\\\text{per unit time}\end{array}\right) = \frac{\left(\begin{array}{c}\text{potential difference}\\\text{between the two points}\end{array}\right)}{\left(\begin{array}{c}\text{resistance to flow}\\\text{between the two points}\end{array}\right)}$$

In a plant, the quantity of water flowing can be calculated from

$$Q = \frac{V}{R_{soil} + R_{root} + R_{stem} + R_{leaves} + R_{vapour\ phase}} \qquad (1.6)$$

$$\left(\begin{array}{c}\text{volume of water flowing}\\\text{per unit time}\end{array}\right) = \frac{\left(\begin{array}{c}\text{potential difference between soil}\\\text{and atmosphere}\end{array}\right)}{\left(\begin{array}{c}\text{sum of the resistances between the}\\\text{soil and atmosphere}\end{array}\right)}$$

In the same way that the resistance term in equation 1.5 has been broken up into its component parts, the potential-difference term can be divided into the potential drop along each part of the water pathway. The sum of the potential drops along each part of the pathway must be equal to the total potential drop along the whole pathway, in exactly the same way as the sum of the series resistances on each part of the pathway is equal to the total resistance from the bulk of the soil to the bulk of the air. A potential drop can occur across any one part, or for that matter across the whole of the pathway, because that part of the pathway constitutes a discrete resistance. If the potential of a leaf fell suddenly and there were no resistance between the leaf and the soil, then all the water needed to equalize the potentials would move virtually instantaneously from the soil to the leaf, and the potentials would again become equal throughout the system. This does not happen because there are resistances to movement through the soil, the plant, and the air (see p. 53).

Water which is ultimately transpired as water vapour in the atmosphere has to flow through the soil, the components of the plant water system, and into the air; at dynamic equilibrium (when flow rates, water potentials and water content do not change) the same amount of water must be moving through each part of the soil-plant-air pathway. As the amount of water that can flow through a part of the pathway is proportional to the potential drop across that part, and inversely proportional to the resistance to flow through it, then

$$Q = \frac{V_{\text{soil}}}{R_{\text{soil}}} = \frac{V_{\text{root}}}{R_{\text{root}}} = \frac{V_{\text{stem}}}{R_{\text{stem}}} = \frac{V_{\text{leaves}}}{R_{\text{leaves}}} = \frac{V_{\text{vapour phase}}}{R_{\text{vapour phase}}} \qquad (1.7)$$

When using equations 1.6 or 1.7, it must be remembered that the potential drops and the resistances are those along all the individual components. Roots, stems and leaves are arranged in series. However, we can consider individual leaves as being in parallel with one another, and likewise individual roots and the xylem elements. Thus the total leaf resistance of a plant with four leaves is given by

$$R_{\text{leaves}} = \frac{1}{\dfrac{1}{R_{\text{leaf 1}}} + \dfrac{1}{R_{\text{leaf 2}}} + \dfrac{1}{R_{\text{leaf 3}}} + \dfrac{1}{R_{\text{leaf 4}}}} \qquad (1.8)$$

The total resistance to flow through the roots could be expressed similarly, substituting *root* for *leaf*; and for the stem, substituting *xylem element* for *leaf* in equation 1.8.

However, when using equations 1.5, 1.6 and 1.7, difficulties arise because the potentials and the resistances should be expressed in the same units for each part of the water path, and this is complicated, as the following discussion will show.

So far, the resistances and potentials have not been defined because water vapour moves along density (or partial-pressure) gradients which can only with difficulty be translated into water potential gradients. Vapour movement follows Fick's Law:

$$v = -D\frac{d\theta}{dx} \qquad \text{kg s}^{-1} \qquad (1.9)$$

where v is the quantity of water vapour (kg s^{-1}) diffusing in unit time; D is the diffusivity, calculated over the area through which diffusion is occurring and therefore of dimensions $(\text{m}^2\,\text{s}^{-1})(\text{m}^2) = \text{m}^4\,\text{s}^{-1}$; and $d\theta/dx$ is the density gradient, i.e. $(\text{kg m}^{-3})(\text{m}^{-1}) = \text{kg m}^{-4}$.

However, liquid water in the plant moves along water potential gradients expressed in bars, which is incompatible with dimensions of water density gradients in g cm^{-3}, especially as densities and potentials are logarithmically related:

$$\psi = \frac{RT}{V_{\text{w}}}\ln\frac{\theta}{\theta_o} \qquad (1.10)$$

where ψ is the water potential, R the gas constant ($8\cdot31\,\text{J K}^{-1}\,\text{mol}^{-1}$),

T is the absolute temperature, θ the water vapour density in the system, and θ_o the water vapour density at $\psi = $ zero.

To resolve the difficulty when using equations 1.6 and 1.7, it will be necessary to use the water potential of the vapour phase as the driving potential to make the vapour phase units compatible with those of the rest of the plant. This can be done by using a valid resistance term which is not constant in value but which changes with changing water vapour density, while being dimensionally equivalent to the other terms in equations 1.6 and 1.7. Such a resistance can be derived from equations 1.9 and 1.10:

$$\text{Resistance}_{\text{gas phase}} = \frac{RT}{DMb} \ln \frac{\theta_s - bl}{\theta_s} \qquad \text{N s m}^{-5} \qquad (1.11)$$

where R is the gas constant, T is the absolute temperature, D is the diffusivity, M is the molecular weight of water, b is the slope of the line "vapour density versus distance from the source", l is the distance over which diffusion occurs, and θ_s is the water vapour density at the source of the diffusing vapour.

When the gaseous phase resistance, as calculated from equation 1.11, and the water potential of the gaseous phase are inserted in equations 1.6 or 1.7, the resulting fraction reduces to Fick's Law, and is therefore valid for water vapour flux in the gaseous phase.

While the "water potential of the gaseous phase" is the best expression to use in certain mathematical analyses, especially when comparing water fluxes in the plant with those in the atmosphere, conceptually it is preferable to use "water vapour density". This term will be used in more generalized discussions in the following chapters.

The measurement of water potential

The only technique which permits the accurate measurement of total water potential depends on the fact that at equilibrium the partial pressure (water potential) of the air above a piece of material, such as a leaf or soil sample, enclosed in a container, is the same as the partial pressure (water potential) of the material itself (see Appendix). This means that if the water potential of the air can be measured by measuring its water vapour content, then at equilibrium the water potential of the material will be known.

Measurements must be made at constant temperature. The vapour content of the air is measured by recording the wet-bulb temperature of a psychrometer, i.e. the parameter *actually* measured is the atmospheric moisture deficit rather than the atmospheric vapour density or the water potential. This is the isopiestic method (p. 116).

The type of psychrometer mostly used to make such measurements is known as the Peltier-cooled psychrometer. If two wires made from different metals are joined by spot welding or solder to form a loop, and the junctions kept at different temperatures, an electric current will flow round the loop. Such a loop is shown in figure 1.2. The magnitude of the current will depend on the temperature difference between the junctions (x) and (y), and on the metals used. This thermoelectric phenomenon is known as the *Seebeck effect*.

Thermocouples can be made to work in reverse by passing a small current through them. When this is done, one junction will be cooled and the other heated. This is the *Peltier effect*, essentially the reverse of the Seebeck effect. In figure 1.3 a Peltier-type thermocouple is shown. The junctions (y') and (y'') must be kept at the same temperature (usually ambient). Although there is a limit to the temperature difference that can be produced between junctions (x) and (y', y'') by passing a current through the assembly, it is usually possible to cool junction (x) sufficiently to cause dew deposition on it. The wires must be very thin (10–25 µm diameter) so that the junctions can change in temperature without a significant effect on the surroundings.

During the psychrometric measurement of water potential, the Peltier-cooled thermocouple is mounted in a small chamber which must be kept at constant temperature ($\pm 0.3\,^\circ$C), and the sample of plant or soil material is enclosed in this chamber. After a sufficiently long time (often several hours) to allow for the liquid and vapour phases to equilibrate, the cooling current is passed in order to cause dew deposition on the junction in the chamber. As the moisture drop that has formed evaporates, a difference in temperature between junctions (x) and (y', y'') will

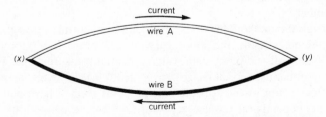

Figure 1.2. Simple loop of dissimilar wires, for the measurement of the Seebeck effect when (x) and (y) are at different temperatures.

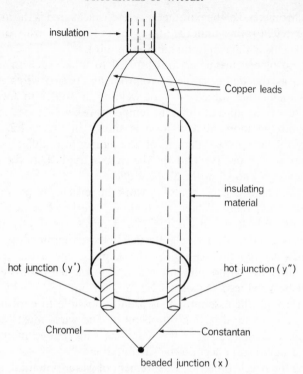

insulation

Copper leads

insulating material

hot junction (y′)

hot junction (y″)

Chromel

Constantan

beaded junction (x)

Figure 1.3. Peltier-type thermocouple for psychrometric measurements in which junction (x) is cooled to dew-point by reverse current flow and "wet" bulb depression is measured as the dew drop evaporates, lowering the temperature at (x) to below that at (y′) and (y″).

develop. The magnitude of this wet-bulb depression is the most accurate measure known of the vapour density in the air above the material in the chamber.

Although the method demands very good temperature control, hardly ever available outside the laboratory, it has been used in the field, especially for soil and plant-stem water-potential measurements. The fine wire junction must, of course, be protected by wire cages or ceramic cups before being inserted in the soil or plant stem. The usefulness of the results depends on the accuracy with which the temperature of the system can be measured, because water vapour deficit is an extremely sensitive function of atmospheric temperature.

1.8. Water and the energy balance

Energy arrives continuously at the surface of the earth, and leaves continuously to be dissipated in space. The energy transfer occurs almost exclusively as a radiant energy flux, i.e. as electromagnetic radiation—more particularly as the type of electromagnetic radiation given off by a body as a function of its temperature, consisting of infra-red radiation, sometimes called *thermal radiation*, and light. Thus there is a continuous radiant-energy flux both towards and away from the earth, usually with a net influx during the day, and a net efflux at night. For practical purposes, it can be said that all the energy arriving at the earth's surface comes originally from the sun, and the energy lost from the earth's surface into space is either in the form of reflected radiation (largely visible light) or as infra-red radiation emitted by the earth. Thus we can talk about the energy balance (the difference between the incoming and outgoing radiation) of the earth's surface, or that of objects on the earth's surface; and about the effect that the energy balance has on other parameters: for example, the energy balance of a leaf or plant will affect its temperature and therefore its water economy. The energy balance of the earth's surface will not be the same, either from one place to the next, or in any one place over a period of time, but will vary both with the thermal nature of the surface (reflectivity, shape,

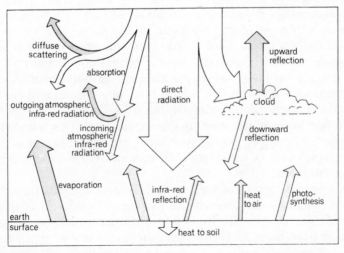

Figure 1.4. Diagrammatic representation of the day-time energy balance at the earth's surface.

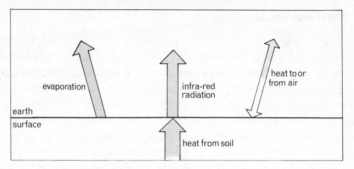

Figure 1.5. Diagrammatic representation of the night-time energy
balance at the earth's surface.

thermal capacity, wetness, etc.) and the time of day. The essentials
of the day-time energy balance at the surface of the earth are summed
up in figure 1.4, and those of the night-time energy balance in figure 1.5.
However, the energy balance at the earth's surface only interests us
here as far as evaporation is concerned, and more particularly in relation
to the energy balance of leaves.

The effect of the energy balance on evaporation from soil or a free
water surface is fairly clear. If the energy input is greater than the
energy loss, the temperature will rise, and evaporation will tend to increase; while if the energy loss is greater, the temperature will fall and
evaporation will be reduced. If evaporation increases by a large amount,
for instance because of an abundance of free water, this will tend to
keep the ambient temperature in and around the water surface constant,
because, as the water evaporates, the energy that would otherwise cause
an ambient temperature rise is absorbed by the water in the form of
latent heat when it evaporates. A large body of water can also moderate
the ambient temperature because it has a high specific heat, and can
absorb heat from, or give it off to, the surroundings. This is known as a
"sensible heat" transfer, because the heat energy is transferred in a form
such that it causes a temperature change, and not a change of state,
as does latent heat.

The energy balance of a leaf, or plant, is more complex than that of
the earth's surface. An exposed leaf receives reflected visible light and
radiated infra-red radiation from the ground underneath, as well as the
radiation from the sun. The day and night-time energy balances of a
single leaf are shown in figures 1.6 and 1.7 respectively. There are three
possible outcomes when radiant energy impinges on a leaf:

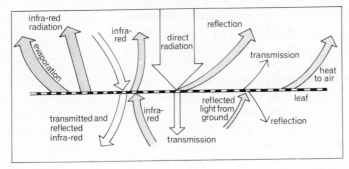

Figure 1.6. Diagrammatic representation of the day-time energy balance of a single leaf.

(1) *Reflection.* Of the incident visible light impinging on a leaf normally less than 20% is reflected, though it is higher in some leaves because of the presence of a light-coloured hairy or waxy layer. The reflectivity of leaves rises sharply in the infra-red region, to a value usually of the order of 40 to 60% of the infra-red received.

(2) *Transmission.* Relatively little of the visible light falling on a leaf is transmitted through it (usually less than 10%), and the proportion passing through decreases with leaf thickness. A larger proportion of the infra-red radiation is transmitted, however, the largest values usually being of the order of 30–40%.

(3) *Absorption.* The visible light and infra-red radiation that is not either reflected or transmitted through the leaf is absorbed (approximately 80% of the visible light, 50% of the total radiation and about 10% of the infra-red). The high absorption of light by leaves that are relatively

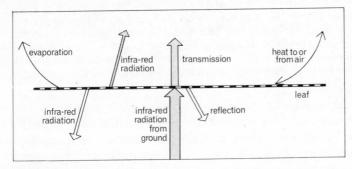

Figure 1.7. Diagrammatic representation of the night-time energy balance of a single leaf.

thin (though absorption generally increases with thickness) is due to backward and forward reflection within the leaf at boundaries of differing optical densities, e.g. at air-water interfaces. This is similar to the multiple reflections that occur in powdered glass, causing it to look white. Of the energy absorbed by a leaf something less than 2% is used on an average for photosynthesis. This means that 98% or more of the energy absorbed by a leaf (frequently over 78% of the total incoming energy, i.e. over $26,110,000\,J\,m^{-2}\,day^{-1}$ in midsummer) is potentially able to raise leaf temperature.

If leaf temperature is not to rise to a point at which cell death occurs, the incoming energy must be dissipated by the leaf. This can occur in a number of ways:

(1) Heat can be removed from a leaf by emission of infra-red radiation. The infra-red radiation emitted by a body is a function of its temperature, but the radiation increases as the fourth power of the absolute temperature (Stefan-Boltzmann law). As the leaf temperature increases, the degree of cooling by radiation also increases, though for radiation impinging on a leaf to be counterbalanced in the daytime by an outward radiation flux, the temperature would have to be so high that tissue damage would occur. At night, however, the much lower energy input could be balanced in this way.

(2) Heat can be carried away by convective cooling, either by:
(a) *Free Convection*. When air is heated it becomes less dense, so that if a part of a mass of air is heated it will rise in relation to the rest of the air. If a leaf is at a higher temperature than the air surrounding it, it will heat the air that is in contact with it and cause it therefore to rise with respect to the rest of the air, thereby setting up a convection current which will cool the leaf.
(b) *Forced Convection*. As already stated, air in contact with an object, e.g. a leaf, that is warmer than the air will be heated. If a wind blows over the object, some of the heated air will be removed by bulk transfer and will be replaced by cooler air from the rest of the atmosphere, which can then be heated. Thus, while there is a wind blowing over the object, and while the object is warmer than air temperature, there will be a continual heat flow away from it.

As far as leaves are concerned, the more important of the two types of convective cooling is usually forced convection, as it is rare for there not to be any wind. However, there are occasions when free and forced convection are of approximately equal importance, i.e. when there is little, or no, air movement and a large difference between the air and leaf temperatures. If a leaf is at a higher temperature than the air, there will not be a sudden temperature discontinuity at the leaf-air interface. The air molecules immediately adjacent to the leaf will be only slightly cooler than the leaf, and the air molecules next to these slightly cooler still. Thus there will be a temperature gradient in the air surrounding the leaf, and even in still air this gradient will be relatively steep, and occur over a short distance of only a few millimetres. The region near the leaf in which this steep temperature gradient occurs is the *boundary layer* (p. 9). In fact the boundary layer as measured for a temperature gradient would be much the same in thickness as that measured for a vapour density gradient on the same leaf at the same time, as the boundary layer is essentially a layer of unstirred air adhering to the leaf.

(3) A considerable amount of heat is dissipated by evaporative cooling, provided the plant has a sufficient water supply to prevent desiccation. Since the stomata must be open to allow carbon dioxide to enter the leaf air-space system, so that it can dissolve in the water films of the mesophyll cell walls, water will be in contact with the internal leaf atmosphere. This leaf atmosphere will be near saturation vapour density, whereas the outside air will be at a lower vapour density, so that vapour diffusion to the outside will occur via the open stomata. Thus a restricted but continuous evaporative cooling is guaranteed as long as the stomata are open.

Experiments

Experiment E1.1. *The tensile strength of water.*

This can be measured by classical experiments using the water hammer (not as simple as it might appear). In the context of this book it is instructive to give an indication only of this property of water by using a leafy twig cut under water with the last 3–5 cm of its bark removed. The twig is fitted into a rubber bung, and the twig together with the bung fitted into a water-filled piece of glass tubing. All operations should be carried out with the cut end of the twig kept under water. The apparatus is then set up as shown in figure E1.1, completely filled with boiled water and without any air bubbles.

Boiled distilled water should be used in the beaker above the clean mercury. On illumination and in a warm gentle breeze the water in the capillary tube will move upwards. After a short while the beaker may be raised so that mercury will begin to ascend the tube. A mercury column of a maximum length of 75 cm can usually be drawn up. When the system breaks, it will do so at the mercury-water interface, but the water column itself will withstand the tension and remain substantially intact because of the cohesion forces within the bulk water and the adhesion forces between the water and glass molecules.

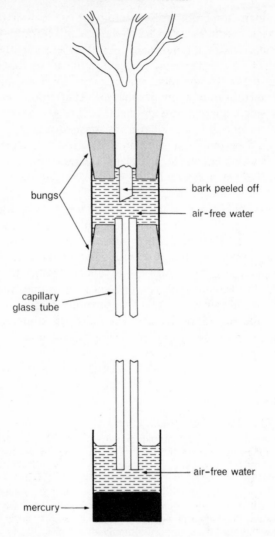

Figure E1.1. Apparatus used for demonstrations of the tensile strength of water. In place of the twig, a line from a vacuum pump could be connected, provided a large capillary resistance is interposed between pump and top reservoir. In this case, however, a maximum of about 75 cm suction can be achieved.

Experiment E1.2. *Volume shrinkage and temperature rise on adsorption.*
Two 250-cm^3 volumetric flasks are used to calibrate a 500-cm^3 flask. Once this has been done, one of the 250-cm^3 flasks is filled with absolute alcohol and the other with distilled water. The temperatures of the liquids in both flasks are measured and recorded. The water is next transferred from the 250-cm^3 to the 500-cm^3 flask using a funnel. This is followed by transferring the alcohol from its flask to the 500-cm^3 flask, using the funnel in such a way that the alcohol splays irregularly on to the water in the flask. It must not be permitted to run down the side of the 500-cm^3 flask. In this manner the alcohol will be seen to form a layer floating on the water, and the two liquids will fill the flask almost to the calibration mark. On quick inversion of the flask, the two liquids will mix instantly, the volume will be seen to have shrunk and, if a thermometer is put into the mixture, its temperature can be measured. A rise of about 10 °C should occur and, if the volume is made up to the calibration mark by adding from a pipette, it will be seen that the volume shrinkage amounts to about 15 cm^3.

Experiment E1.3. *Energy release on adsorption.*
Half a level teaspoonful of previously dried potato starch is placed in a 50-cm^3 beaker. The beaker should be held at its rim and slightly inclined, so that the starch collects at one side. Using a thermometer, the starch is gently stirred and its temperature is measured. The temperature of the water should also be measured, and the average temperature of starch and water mark the starting-point. Three drops of distilled water are allowed to fall from a pipette into the starch, while the mixture is gently stirred with a thermometer. The temperature should be read within a few seconds and recorded. This is repeated several times, as speedily as possible and until the temperature begins to decrease.

By plotting the changes in temperature of the mixture against the number of drops of water added, a curve which rises steeply at first, then gradually flattens out and finally declines, should be obtained. A rise of about 10 °C should be achieved; it is due to the release of kinetic energy from water molecules when they are restricted in their kinetic movement by being adsorbed on to colloidal starch.

Experiment E1.4. *Comparative method of measuring surface tensions.*
Ten drops of test liquids, such as‒distilled water with a trace of detergent, alcohol, turpentine oil or other natural products, as well as distilled water alone, which is used as a standard, are allowed to fall from graduated 1-cm^3 pipettes, and the volumes of the ten drops are read accurately to the third decimal place. Pipettes should be held vertically and, for each liquid, at least three replicate measurements of volumes should be carried out. Surface tensions are calculated according to the following expression:

$$\text{S.T.} = \frac{10^{-3} \times 73 \times B \times \text{S.G.}}{A} \qquad \text{N m}^{-1}$$

A = volume of 10 drops of water.
B = volume of 10 drops of test liquid.
S.G. = specific gravity of test solution.

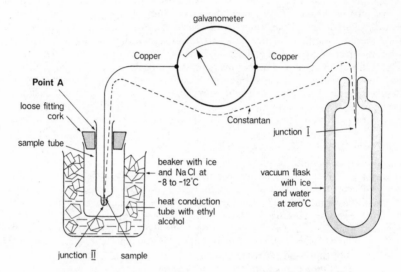

Figure E1.2. Apparatus suitable for cryoscopy and the demonstration of the release of the latent heat of fusion when water "sets" to ice (p. 120).

Experiment E1.5. *Release of the latent heat of fusion.*

The simple cryoscopic apparatus shown in figure E1.2 is assembled with a small drop of clean distilled water in the sample tube. Supercooling by 1 °C, followed by a sharp but gentle knock at point A, will cause the droplet to "set" instantly into ice, and this is accompanied by a dramatic rise in temperature to the true freezing-point. The supercooling process should be "smooth", i.e. the galvanometer needle should move evenly at a rate soon appreciated by experience; if this does not happen, the cause is probably dirty water, which freezes as the temperature falls—or, in other words, cannot be supercooled.

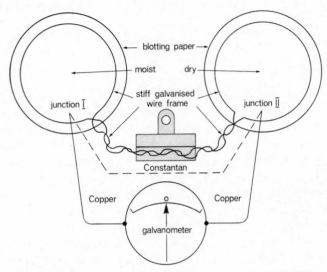

Figure E1.3. Apparatus used for the measurement of the cooling effect of evaporation due to the absorption of the latent heat of vaporization.

Experiment E1.6. *The cooling effect of evaporation.*

Two pieces of blotting paper glued on to wire frames are assembled as shown in figure E1.3 and connected to the thermocouple circuit as shown. One paper is wetted with water one or two degrees above room temperature; this will cause a small temperature differential, which is recorded. A slow fan is now directed at the two pieces of blotting paper so that the air about them is evenly stirred. The temperature differential will grow considerably with the moist paper becoming cooler by at least 4 °C at a room temperature of about 20 °C.

CHAPTER TWO

WATER VAPOUR AND THE ATMOSPHERE

2.1. Evaporation

Water vapour will diffuse away from a water surface exposed to an unsaturated atmosphere at a rate which depends on the atmospheric water vapour density, the resistance of the path between the water surface and the bulk atmosphere, and the water potential of the source. In addition, temperature and wind greatly influence the rate of evaporation.

Temperature. The effect of temperature on evaporation from a free water surface is complex, because the liquid and the vapour are often at different temperatures. A rise in the temperature of the atmosphere will cause both an increase in its vapour-holding capacity, and a decrease in the diffusive resistances (p. 9); the latter is at least partially due to the correlated increase in the kinetic energy of the molecules in the vapour diffusion path. The effect of a rise in temperature in the liquid phase is difficult to predict under natural conditions, because it is rare for the liquid-phase temperature to increase without a simultaneous, though often different, increase in the vapour temperature. However, the basic effect will be one of increasing the mean kinetic energy of the liquid water molecules, so that more of them will have sufficient momentum to overcome the forces holding them in the liquid (p. 3) and they will escape at an increased rate. The effects of temperature on evaporation are complex also because of the absorption of latent heat ($2441 \, \mathrm{J \, g^{-1}}$ at $25\,^{\circ}\mathrm{C}$). This heat must come from the surroundings and tends to cause the temperature of the system to decrease.

Wind. A linear increase in wind speed over a free water surface will cause an approximately exponential increase in evaporation, as it will cause an approximately exponential decrease in the diffusive resistances, largely affecting the boundary layer (chapter 1), and it may also change the atmospheric vapour density. Again there will be a cooling effect, though at the same time there will be a heat input from the approaching air which, together with the radiant energy input from the sun during the day, may more than compensate the heat lost due to cooling; during the night, however, the cooling effect may well be measurable.

Potential evaporation rate is the maximum evaporation rate from a free water surface at ground level, and is most accurately measured with sunken evaporimeter pans. These are large metal dishes, about 30 cm in diameter, with their rim at ground level and kept automatically filled to the brim with water; the rate of water loss from such pans is taken as the potential evaporation rate at that place and under the prevailing conditions. Other methods for measuring or estimating the potential evaporation rate from climatic data are described in Kozlowski's *Water Deficits and Plant Growth*, Vol. 1 (1968).

2.2. Vapour densities in leaf air spaces

The water vapour density in air spaces of leaves with their stomata closed is at all times near saturation density at the prevailing *leaf* temperature. Under these conditions the potential of the water in the cell walls will be close to zero, as will the water potential of the mesophyll cells with which this water will be in equilibrium.

When the stomata open and vapour diffusion to the outside begins, a gradient from 100% saturation at the sites of evaporation, i.e. the moist cell walls, to the stomatal cavities must develop as diffusion proceeds. In the stomatal cavity itself, the density does not normally decrease to less than about 96% of saturation, at which point the leaf water potential will be in the region of −50 bar (p. 41). In place of a state of static equilibrium, there will now be a dynamic gradient in water potentials between the water in the cell walls and the vapour phase. As soon as the density departs from 100% saturation (p. 8), the potential of the vapour will be lower than that of the liquid phase, e.g. by 13·7 bar at 99% of saturation and by 27·5 bar at 98% of saturation. This leaves out of account other gradients between the vascular tissue and the inner epidermal and mesophyll walls (chapter 3). The potential of the water in the cell walls (mainly due to matric forces) will decrease as water is lost by evaporation and the menisci in the pores retreat into narrower capillaries. This causes a reduction in pressure of the water in the walls, which is represented by the expression:

$$P_{cap} = 2 \text{ S.T.} \left(\frac{1}{r_2} - \frac{1}{r_1} \right)$$

where S.T. is the surface tension of water and r_1, r_2 are the radii of curvature of the menisci before and after their retreating. However, as long as the water potential of the cells ($\psi_{cell} = \psi_p + \psi_\pi$) remains reasonably high, water will flow into the cell walls because a difference in water potential will be maintained. Even when the water potential of the cells decreases rather drastically due to loss of turgor (ψ_p) and decreased osmotic potentials (ψ_π), the continuous water system in the cell walls which extends throughout the xylem and to the soil water films (chapter 3) will supply water to mesophyll cells as well as to evaporation sites, both supplies being due to gradients in water potentials. The gradient from liquid water in cell walls to water vapour in the atmosphere is usually the steeper of the two, unless the "air entry value" has been reached (p. 65).

The degree of saturation maintained in leaf air spaces when stomata are open depends, as indicated above, on the rate of outward diffusion of vapour from the leaf, and this in turn is determined by stomatal resistances and atmospheric vapour densities. It may be mentioned here that the potential evaporation rate (p. 27) inside a leaf is appreciably higher than that from an open water surface of the same area as the leaf, because the total area from which evaporation occurs (the inner epidermal and mesophyll walls) may be from seven to thirty times the area of the leaf. Nevertheless, the leaf should be considered as an organ serving to retain water in the plant (p. 36) for the following reasons.

It would seem that the inner epidermal wall in close proximity to the stomatal pore is a major evaporation site within the leaf, because diffusion paths of least resistance will enable vapour originating there to escape most speedily; this is indicated in figure 2.1. A relatively larger vapour density difference would be created at these sites, speeding up evaporation as compared to the mesophyll walls which are more distant from the stomatal pore. The epidermal tissue has also been found to have a relatively high hydraulic conductivity (chapter 3), so that the replenishment of water lost by evaporation is assured. Thus the major air space volume among the mesophyll cells remains near saturation and vapour loss from mesophyll walls will be comparatively slow, serving to retain liquid water in the walls. In addition, the hydraulic conductivity (pp. 43, 54) to water movement in the outer part of mesophyll cell walls decreases with developing water stress, preventing too drastic

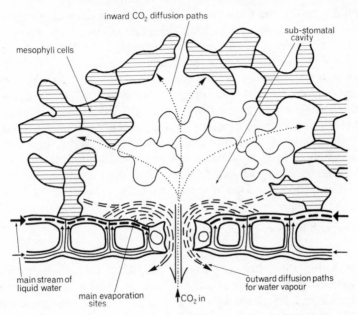

inward CO_2 diffusion paths

sub-stomatal cavity

mesophyll cells

main stream of liquid water

main evaporation sites

CO_2 in

outward diffusion paths for water vapour

Figure 2.1. Diagrammatic representation of vapour flow lines between leaf air spaces and the outside atmosphere. The area of inner epidermal wall accounts for a third of the total internal cavity surface.

a fall in the degree of saturation of the leaf air-space system. Values of less than 96% of saturation occur rarely (see p. 27).

Another factor which has an important bearing on the vapour density in leaf air spaces is leaf temperature. When exposed to high radiation loads, leaf temperatures may rise considerably above the ambient value; in warm climates leaf temperatures of ambient plus 15 °C have often been measured, and increases of between 5 and 8 °C are common in temperate climates. These increased leaf temperatures occur in spite of the high specific heat of water which constitutes between 80 and 90% of the fresh weight of most leaves and which permits a considerable uptake of radiant energy without great increases in temperature. They also occur in spite of the cooling effect of evaporation referred to on pp. 5, 21, 25. This emphasizes that in the absence of this cooling effect leaf temperatures would more often reach lethal levels than we experience in nature.

On rare occasions it can happen that the density of water vapour in the atmosphere is greater than that in leaves. This will occur when plants are under considerable water stress and the atmosphere is near its saturation density. Such conditions exist in arid regions at night, when the soil is sufficiently dry to keep the plant under water stress even at night, and the air temperature falls to the dew point. Water movement into the plant will then occur, especially if the stomata are open; in plants with Crassulacean acid metabolism (CAM, p. 133) which grow in such regions, stomata do open at night and often close during the day, while many non-CAM plants will open their stomata at night if the atmospheric humidity is high. Condensation of vapour on cell walls bordering leaf air spaces could then occur, and inward translocation of such condensed water would be possible, while dew droplets deposited on the surface of leaves are often absorbed, frequently through hydathodes (p. 114) or specialized cells.

2.3. The path of water vapour out of leaves

Leaves function as gas exchange organs by virtue of the aeration system within their structure (figure 2.2). Thus the air-space system in a tobacco leaf occupies about 40 per cent of its total volume, that of a barley leaf 30%, that of a pine needle 5%, and that of a maize leaf 10%. Carbon dioxide from the atmosphere has therefore access to the films of water associated with mesophyll cell walls, and the water vapour present in the air-space systems will have access to the atmosphere when the stomatal pores are open. We have seen in sections 1.6 and 2.2 that there must be water vapour present in the air spaces of a leaf, because the vapour phase in contact with the liquid phase must be in equilibrium with it.

In many leaves the extent of the air-space system changes with changes in leaf water content, so that the quoted percentages must not be taken as constants. In fact, the air-space volume tends to change proportionately more than the total leaf volume when water stress develops. This produces increases in the internal diffusion resistances especially affecting carbon dioxide.

Leaf Diffusion Resistance. There is only one path for water *vapour* diffusion out of a leaf and that is via the stomata when these permit vapour passage through their pores (cf. p. 34 on "cuticular transpiration"). The vapour diffusion path is a composite one consisting of portions of the leaf air-space system itself, the entrance into, passage through, and exit from, the stomatal pores, and then the boundary layer adhering to the leaf as a whole; these paths are all arranged in series. Col-

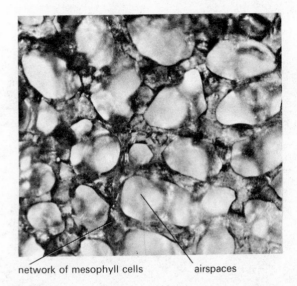

network of mesophyll cells   airspaces

Figure 2.2 Mesophyll cells and associated air spaces below the epidermis of a leaf of *Vicia faba*. Surface view after removal of the lower epidermis. If the upper epidermis is removed, a very similar picture is seen. The main difference is that the palisade cells appear in oval and round cross-sections, but forming a network similar to the spongy mesophyll cells shown in this photograph.

lectively, the resistance of this composite path is called *leaf diffusion resistance*.

The leaf diffusion resistance to vapour flux depends on the geometry of the leaf, i.e. its cross-sectional area and its length. The most outstanding feature of this resistance is its variability. This is due to changes in external factors, such as the occurrence of air movements affecting the boundary layer resistance, and changes in internal factors, such as the air-space system itself, when leaf water content and leaf thickness changes occur. Decreases of 1% in leaf water content cause between 4 and 7% decreases in thickness and up to 10 or 12% in air-space volumes. However, changes in stomatal openings are the most important factor affecting leaf diffusion resistance (figure 2.3).

Stomatal physiology is a subject for study in its own right and, since it is of basic importance to leaf water relations, we must give a summary of some relevant facts. Movements of stomatal guard cells result when the turgor relations between epidermal and guard cells

terminal subsidiary
cell

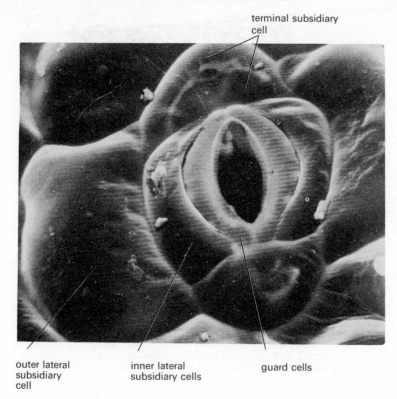

outer lateral   inner lateral   guard cells
subsidiary   subsidiary cells
cell

Figure 2.3. Scanning electron micrograph of an open stoma in an epidermal strip of *Commelina communis*.

change. Such changes occur as a result of water movements following either guard-cell metabolic activity or the influence of environmental factors such as atmospheric vapour density, wind, leaf and air temperatures on the rates of vapour loss. If, in consequence, there occur differential rates of water supply to, or vapour loss from, epidermal and guard cells, these will speedily affect their turgor relations and hence initiate stomatal movements which are directly related to cell turgor relations. Such movements are referred to as *transient* because the equilibrium degree of stomatal opening eventually established will be determined by guard cell metabolism. For instance, illumination at dawn stimulates photosynthetic activity and ionic transport mechanisms leading to changes in osmotic potentials and hence turgor relations which eventually establish new equilibrium stomatal openings. Whilst the role of ionic transport into and out of guard cells appears to be straight-forward in influencing osmotic potentials, the role of photosynthesis is not easily identified. It inevitably leads to a reduction of the concentration of carbon dioxide in the leaf tissue, and this in itself, even if experimentally brought about in the dark, greatly enhances stomatal opening. Whatever the metabolic mechanisms are that determine equilibrium stomatal opening, the transient movements, however incipient, may well precede those due to metabolic processes under natural conditions, when at sunrise leaf temperatures and atmospheric vapour densities change. It might also be mentioned that in many species the stomata are subject to an endogenous rhythm which brings the guard cells into a state of preparedness to open at about dawn.

In the quantitative estimation of resistances of a diffusion path consisting of different and irregular cross-sectional areas, resistances are expressed in equivalent lengths of tubes of unit cross-sectional area which would offer the same resistance to diffusive flow. Thus the geometric dimensions of resistances are metres, i.e. lengths of tubes of unit cross-sectional area; they are converted into diffusive resistances on being divided by the diffusion coefficient of water vapour (p. 9) and have then the dimensions: $s\,m^{-1}$ or $s\,cm^{-1}$. (A detailed account of the estimation of water vapour diffusion resistances can be found in Meidner and Mansfield, 1968). (Expt. E2.5.)

The variability of the leaf diffusive resistance to vapour flux referred to above makes it a regulatory mechanism (p. 37) for the rate of water vapour diffusion, because the rate of diffusion is directly proportional to the water vapour density difference between the leaf air space and the atmosphere, and inversely proportional to the resistance offered by the path. The resistance itself is proportional to the length of the path and inversely proportional to its cross-sectional area.

$$\text{Rate} = \frac{\text{mass}}{\text{time}} = \frac{\text{diffusion coefficient} \times \text{density difference}}{\text{resistance per unit leaf area}}$$

$$= \frac{D \times \Delta\rho}{R}$$

$$= \frac{(m^2\,s^{-1})\,(\mu g\,m^{-3})}{m \qquad m^{-2}}$$

$$= \mu g\,s^{-1}$$

Estimates of the leaf air-space resistance are somewhat arbitrary; the length of the path is usually taken to be half the leaf thickness (L), and its cross-sectional area either that of the leaf blade as a whole, or more commonly estimated per m^2. A typical value for a 600-μm thick leaf is $0.11 \, s \, cm^{-1}$.

The stomatal resistance can be estimated from microscopic measurements of pore dimensions; it varies between $0.8 \, s \, cm^{-1}$ to $7.0 \, s \, cm^{-1}$.

$$R_{\text{stomata}} = \frac{A(1 + \frac{1}{4}\pi d)}{\mathbf{D} n a} \quad \begin{array}{l} \text{for stomatal openings} \\ \text{found under natural} \\ \text{conditions} \end{array}$$

where A = leaf area in cm^2, l = length of stomatal tube in cm, d = diameter of stomatal pore in cm, n = number of stomata in cm^{-2}, a = area of stomatal pore in cm^2, $\mathbf{D}$ = diffusion coefficient of vapour in $cm^2 \, s^{-1}$.

The boundary layer resistance is again a somewhat arbitrary estimate; its cross-sectional area is that of the blade or, more commonly, simply estimated per cm^2, as in the case of the leaf air-space resistance. Its length varies with wind speed. At $1.0 \, m \, s^{-1}$ wind speed the boundary layer would be 1 mm thick, offering a resistance of about $1.0 \, s \, cm^{-1}$, at $10 \, m \, s^{-1} \, 0.35 \, s \, cm^{-1}$, and at $0.1 \, m \, s^{-1} \, 3.0 \, s \, cm^{-1}$.

The expression:

$$r_a = \frac{0.9 \, D^{0.46}}{V^{0.56}}$$

has been derived to give realistic values for boundary-layer resistances in windy conditions. r_a = boundary layer resistance, D = the diameter of the leaf in metres, V = the wind speed in $m \, s^{-1}$.

2.4. Transpiration

The word *transpiration* refers to the loss of water vapour from plants. This loss is a diffusion process and occurs via two paths in parallel: the stomatal and the cuticular path. Of the two, the stomatal is quantitatively the more important. It differs also from the cuticular in that the cuticular "path" is in reality a liquid-vapour interface at which evaporation occurs, whereas the stomatal path is a structural path for vapour movement interposed between an air space already filled with vapour and the atmosphere.

(*a*) *Cuticular.* The cuticle of leaves is a complex extra-cellular structure closely adhering to outer cellulose cell walls of the epidermis. It is not a continuous layer of waxy substances, but a discontinuous assembly of

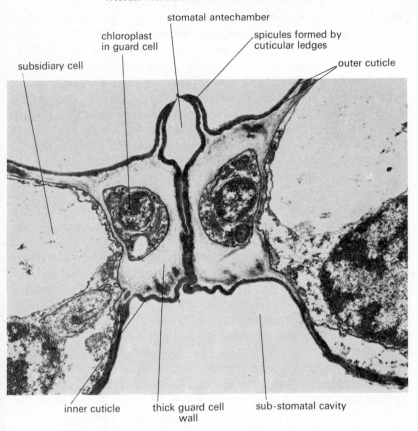

stomatal antechamber

chloroplast
in guard cell

spicules formed by
cuticular ledges

subsidiary cell

outer cuticle

inner cuticle

thick guard cell
wall

sub-stomatal cavity

Figure 2.4. Electron micrograph of stomatal guard cells in transverse
section showing the spicules and continuous cuticle from the leaf
surface into the stomatal cavity (*Zea mays*).

more or less discrete fragments of different chemical composition, which
form the cuticular layer. This layer varies considerably between species,
between leaves of different ages, between upper and lower surfaces of
the same leaf, and between leaves grown in the open and in a glasshouse
or indoors. In all cases, however, the cuticle can vary in the amount of
water it contains at different times. Before sunrise cuticles are often
swollen and comparatively full of moisture. This moisture can originate
externally from dew or internally from the water that saturates the
pecto-cellulosic walls of the epidermal cells.

After sunrise it is the cuticular moisture which is lost to the atmosphere in the form of vapour when the rays of the sun raise leaf and air temperatures ever so slightly. With the evaporation of cuticular water the cuticle may become measurably thinner, its water content lower, and the surface area of water held in the cuticle reduced. This water surface is in contact with the air and is the actual evaporation site. Moreover, the resistance to liquid water movement into the cuticle may also increase so that, as cuticular moisture is lost, it will be replenished at a reduced rate from the reservoir of liquid water saturating the epidermal cell walls. Thus cuticular transpiration will proceed at a rate which depends on the atmospheric water vapour deficit, the extent of the area of water surface in the cuticle exposed to the air, and the rate with which water evaporating at this surface can be replenished from within the leaf.

Reliable measurements of cuticular diffusion resistances are not easily found; they can only be obtained from non-stomatal epidermes, because it is never certain that the stomata are completely closed. Indeed, only cobalt chloride paper measurements (Meidner and Mansfield, 1968) or those obtained with diffusion porometers can be used for measurements of cuticular resistance; and the latter, though known to F. Darwin in 1898, have come into common use only during the past few years. Values for cuticular resistances of non-stomatal epidermes measured with a specially sensitive diffusion porometer vary between a minimum of $40 \, s \, cm^{-1}$ and a still measurable maximum of between 80 and $120 \, s \, cm^{-1}$. Generally, lower resistances are measured when stomatal transpiration proceeds at a low rate and the leaves are shaded. Growing conditions and age influence cuticular resistances within species. Young leaves have higher cuticular resistances than older leaves of the same species because their cuticle is more waxy, and it almost invariably suffers physical damage when the leaves are blown about in the wind or are otherwise knocked.

Compared with the diffusion resistance of stomata, the cuticular resistance is very high indeed, and its contribution to total leaf diffusion resistance therefore small as it represents the resistance of a path in parallel.

$$\frac{1}{R_{leaf}} = \frac{1}{R_{stomata}} + \frac{1}{R_{cuticle}}$$

(b) Stomatal. Although stomatal and cuticular transpiration proceed through parallel paths, there are physiological connections between the two, e.g. stomatal opening is the result of changes in the turgor relations between epidermal and guard cells. When, after sunrise, opening occurs in response to illumination, the epidermal cells may loose some of their turgor owing to cuticular transpiration. This will speed up stomatal opening. Another connection is that the boundary layer resistance, especially important in conditions of little air movement, can be thought of as due, in the first place, to cuticular transpiration, whilst stomata are still closed in the morning; when they open, this existing boundary layer will offer a resistance, slowing down the early stages of stomatal transpiration. However, the situation will quickly be reversed, and the stronger stomatal transpiration will soon contribute a major share to the boundary resistance, which will affect both cuticular and stomatal transpiration alike.

One of the most remarkable features of stomatal transpiration is that, in the case of a leaf with an entire margin, it amounts to about 90% of the evaporation rate from an open water surface of the same size and similar shape as the leaf. Yet the collective total pore area of open stomata amounts to only 1% to 2% of the leaf area. For this reason stomata have been described as "efficient" paths for vapour diffusion. This efficiency is due to stomatal pore size and their distribution in the epidermis. Maximum pore sizes vary from $3\,\mu m \times 10\,\mu m$ to $6\,\mu m \times 40\,\mu m$, and numbers from $10\,mm^{-2}$ to $370\,mm^{-2}$. The efficiency of small-size pores as diffusion paths can be ascribed to the fact that molecules diffusing at the perimeter of the pore can speedily diffuse away from the edge once they are through the pore, provided neighbouring pores are not too close. Hence the importance of the distribution of stomata in the epidermis. As a rule stomata are spaced in such a way that the distance between pore centres is not much less than ten times the maximum diameter of the open pore. This is thought to allow for the relatively speedy "edge diffusion", proportionately the more important the smaller the pore because geometrically the ratio of perimeter to area increases with decreasing pore size. Many small pores are *relatively* more efficient than a few larger ones, always provided they are not too close together. (For detailed discussions of these topics see Meidner and Mansfield, 1968.)

As mentioned earlier, the variability of leaf diffusion resistances (p. 33) makes them regulatory mechanisms. Although all three components of leaf diffusion resistance are variable, it is the significance of changes in the stomatal component which deserves closer consideration. Closed stomata can be considered to offer infinite resistance to diffusion of water vapour. When fully open this resistance may reduce to $0.5\,s\,cm^{-1}$;

in other words, stomatal function can virtually prevent vapour loss, or facilitate it at near maximum rates for the prevailing temperature and moisture deficit conditions.

As we said on page 4, this book really deals with the mechanisms which have evolved to ensure a supply of moisture to, and its retention in, plants. The siting of variable stomatal resistances between the vapour phase inside the leaf and that in the atmosphere is of the highest significance. Nowhere else in the plant could so effective a mechanism for water retention be situated. In the liquid phase, such a variable resistance could not ensure moisture retention in the plant because the steepest gradient in water potential is *out* of the plant across the stomatal pores; at the evaporation sites between liquid and vapour phases the water potential difference is much smaller. A variable resistance here could ensure moisture retention in the plant, but it would be at the cost of the occurrence of exposed water films required for photosynthesis.

The only effective situation for such a variable resistance is where we find it, namely in the epidermis in the form of stomata interposed between vapour-filled air spaces and the atmosphere. Such a variable resistance here can ensure moisture retention in the plant, and moreover ensure that there will be liquid water films for carbon dioxide absorption, as well as a reduced rate of evaporation from these liquid water films. This is because the air adjacent to them in the leaf air-space system will remain near saturation.

2.5. Rates of transpiration

Although the loss of water vapour from leaves occurs both via the cuticle and via the stomata, we confine the discussion of rates to stomatal transpiration because of the paucity of reliable data about cuticular transpiration. We merely add to the comments made in section 2.4 that in most species cuticular transpiration is thought to account rarely for more than 3% of total transpiration, though higher values (10–30%) have been estimated by some workers. The evaporative factors discussed below and used quantitatively in calculations of rates of stomatal transpiration are valid, of course, also for cuticular transpiration. The appropriate value for the cuticular resistance (40–80 s cm^{-1}) can be substituted to obtain comparative rates of cuticular transpiration.

Factors Affecting Transpiration Rates. Rates of stomatal transpiration are primarily determined by the interplay between degree of stomatal opening, leaf and air temperature, and atmospheric water vapour deficit.

Stomatal Resistance. As discussed in section 2.3, under conditions of mild air movements and over a certain range of opening, stomatal diffusive resistance is operative in controlling plant-water status, because of its decisive influence on rates of vapour diffusion out of leaves. Stomatal responses to environmental factors are not the subject of this text, and it is merely stated that stomata respond to a variety of interacting factors of which light, carbon dioxide concentration, temperature, wind, leaf water content and atmospheric moisture content are the most important. In addition, stomata are subject to an endogenous rhythm in opening and closing movements (for details on stomata see Meidner and Mansfield, 1968).

Only the direct physical effect of light on leaves must be looked at more closely. Under natural conditions, leaves become illuminated at the time when their stomata are in a condition of preparedness to open due to their endogenous rhythm. Thus, some degree of stomatal opening will occur shortly after sunrise. However, in most plants it is only the stomata of leaves facing east and south-east which are illuminated fully and therefore continue their opening movement to a maximum as the sun rises. In others, the opening movement is interrupted temporarily, and continues only when the rays of the sun fall fully on the leaf during the course of the day. In many leaves gradual and partial stomatal closure sets in as the leaves move into the shade with the passage of the sun. This should not be confused with the so-called phenomenon of "midday-closure" which is mentioned below as a consequence of either higher leaf temperature or water stress.

Leaf and Air Temperature and Atmospheric Moisture Deficit. One consequence of a leaf being illuminated is a rise in its temperature, and the inevitable increase in the saturation water vapour density inside the leaf air spaces at the very time when the stomata begin to open or continue to open more fully. The cooling effect of transpiration (pp. 25, 29) may counteract this increase in leaf temperature, but as a rule it does not annul it. At the same time the rays of the sun begin to warm up the air and inevitably enlarge the atmospheric water vapour *deficit*. Thus, as shown by the figures in Table 2.1, the difference in water vapour pressure between the leaf and the atmosphere increases. For instance, if both a leaf and the air are at $16\,°C$ and the air is 75% saturated, the water vapour

Table 2.1. Saturated vapour pressures and vapour pressure deficits at different percentage of saturation for two temperatures

Temperature	Vapour pressure		Vapour pressure deficit
°C	% of saturation	mbar	mbar
15	100	17·0	zero
	75	12·8	4·2
	70	11·9	5·1
	60	10·2	6·8
	50	8·5	8·5
20	100	23·3	zero
	75	17·5	5·8
	70	16·3	7·0
	60	14·0	9·3
	50	11·6	11·6

It should be realized that with a rise in temperature without any change in the absolute amount of vapour in the atmosphere, the vapour pressure DEFICIT increases.

pressure difference will be 4·5 mbar. If both leaf and air temperatures rise to 18 °C, and the humidity of the atmosphere decreases to 70% of saturation, this difference will have increased significantly to 6·1 mbar, but if the warming effect is differential, so that the leaf reaches 23 °C and the air stays at 18 °C and at 70% of saturation, the vapour pressure difference increases considerably to 13·6 mbar. These values are for temperate climates. In warmer climates the effects are much more pronounced (Expt. E2.3).

Air Movements. The effects of air movements on the rate of transpiration are due to their effects on the boundary layer resistance (p. 34); they are also a function of leaf surface features such as hairs. In section 2.3 some values for boundary layer resistances were cited. It is, however, the interaction between boundary layer resistance and stomatal resistance that is of interest in its influence on transpiration rates. Bange's well-known graph reproduced in Meidner and Mansfield (1968) is a good illustration of this interaction. The data show that in moving air of about $3 \, \text{m s}^{-1}$ the degree of stomatal opening controls the rate of transpiration over the whole range of opening, somewhat diminished at the larger openings. In still air stomatal control is less pronounced and diminishes more sharply once stomata are well open, the absolute rate reached never approaching that in moving air because of the large boundary-layer resistance. This graph has been reproduced often. It is therefore not out of place to point out that "still" air hardly ever exists in nature, and stomatal opening in *Zebrina p.* under natural conditions rarely exceeds 12 μm. Thus, under natural conditions, pronounced stomatal control of transpiration rate is found practically over the whole range of stomatal opening.

Leaf Water Content and Air-space Resistance. It is almost certain that under temperate climatic conditions leaf water content affects rates of transpiration only indirectly via its effect on the degree of stomatal opening (and via the amount of water held in the cuticle). If the leaf water content falls from e.g. 500% of dry weight at full turgidity to 460% of dry weight prior to wilting, it can still be assumed that liquid water films are present in internal cell walls and that evaporation into leaf air spaces will keep these at near saturation water vapour pressure. It must be realized, however, that if net diffusive flow of water vapour to the outside occurs, there must exist a gradient in water vapour density from near saturation next to the liquid water film to a value significantly less in the sub-stomatal cavity. A humidity of 99·8% of saturation corresponds to a water potential of $-2\cdot4$ bar, but at a humidity of 96% of saturation the water potential is $-55\cdot8$ bar at 20 °C.

Following net diffusion of vapour to the outside, rates of liquid water movement into the leaf will rarely match the rates of vapour loss, and hence some degree of water stress will develop in the leaf—indeed, it must develop if liquid water movement down the water potential gradient is to occur (chapter 3). However, such water stress does not appear to affect seriously either the rate of evaporation due to a postulated lowering of the vapour density (see below), or the extent of liquid water surface inside the cellulose cell walls in contact with the air space.

Mild degrees of water stress do affect the turgidity of leaf cells, especially of the epidermis. Hence there can be an effect of water stress on the degree of stomatal opening, because the latter depends on the turgor pressure differential between epidermal and guard cells (p. 32). But it should not be thought that water stress must impair stomatal functioning. Indeed, it is almost certain that there is an optimum water deficit for the stomatal mechanism which would seem to coincide with the loss of most of their turgor pressure by the epidermal cells. An adverse effect of water stress on stomata begins to occur, as a rule, when the leaf water potential has fallen to between -12 and -15 bar, i.e. when the difference between ψ_{leaf} and ψ_{π} of guard cells no longer allows for sufficient guard cell turgor to be maintained. Stomatal response to water stress can be threefold, depending on the *rate* with which water stress develops and on its magnitude.

1. If stress develops gradually, guard cells will be unable to obtain sufficient water from neighbouring cells and cell walls to maintain their turgor, and partial closure will set in; this will reduce rates of transpirational water vapour loss, so that an equilibrium state might be achieved when rates of leaf water supply and rates of transpiration will be matched.

2. If stress develops rather suddenly, and during a time of vigorous transpiration, the sudden loss of epidermal turgor can result in "transient opening of stomata". In this situation subsidiary and epidermal cells lose their turgor, while the guard cells with their low osmotic potentials can maintain their turgor for a few minutes; therefore they are able to push the surrounding epidermal cells out of place and temporarily open the pore further. Transpiration will then proceed even more vigorously than before, and the leaf will soon begin to wilt with subsequent complete stomatal closure. This is relevant to certain phenomena discussed in chapter 3. Transpiration having then practically stopped, recovery from wilting can take place.

3. The third kind of response to stress is a hormone-initiated partial closure which can be short-lived or extend over several days gradually diminishing. During water stress the hormone abscisic acid accumulates in guard cells and causes partial or full closure; concentrations of 10^{-8} molar are effective.

The Water Potential at the Evaporation Surface. When the leaf water potential is high or the transpiration relatively low, the humidity of the air space in the leaf does not deviate significantly from saturation. However, if the water potential of the leaf tissue falls substantially, or the transpiration rate becomes high, the humidity of the leaf air space may decrease considerably.

Retreat of Menisci. As the water potential of the leaf tissue falls, the menisci in the pores of the micro-fibril mesh of the cell walls will tend to be drawn inwards (retreat) from the wall surface. Thus, in addition to the decrease in humidity due to the lower water potential in the leaf, there will be an extra vapour phase resistance introduced as the length and tortuosity of the vapour pathway increases. This phenomenon will only occur when plants become seriously stressed, because the pores in the wall mesh are very small. In fact water potentials of -150 bar to −300 bar would be required to drain interfibrillar spaces of the sizes commonly found (5–10 µm diameter), while a water potential of −15 bar would be needed to drain pores of 100 µm diameter; and pores of this size probably occur rarely.

Hydraulic Resistance. The outer surfaces of the mesophyll cell walls are hydrophobic because they are cutinized. The outer layer of cell walls has a higher hydraulic resistance than the rest of the liquid pathway in the walls, so that especially at high transpiration rates the water potentials of the outer cell wall surfaces may decrease below those of the rest of the leaf. This causes a fall in the humidity in the leaf air spaces, which has been shown to occur in cotton leaves, from a relative vapour pressure of 1 at zero transpiration, to 0·7 at a transpiration rate approximately $3·6 \, cm^3 \, dm^{-2} \, h^{-1}$. The hydraulic resistance of the outer wall increases with both increasing transpiration rate and de-hydration. This is probably due to changes in the character of the pathway during dehydration; thus it is known that the resistance of cutin to water movement through it increases as it dehydrates. However, it may also be due to a decrease in the area from which evaporation occurs.

Solutes. The osmotic potential of xylem sap is usually of the order of −1 to −2 bar, and it might therefore be expected that as pure water is being removed from the evaporation surface in the leaf, solutes would accumulate at the evaporation sites. Such an accumulation would lower the osmotic potential, and therefore the water potential, on the cell-wall surfaces and hence cause the vapour density in the leaf air spaces to fall. This could occur transiently, for a period of up to one or two minutes, if the transpiration rate increased suddenly and dramatically, but re-equilibrium would soon occur by back-diffusion of the solutes. The osmotic potentials at the evaporation surfaces may at most be 15% lower than those of the xylem sap, even under extreme trans-piration rates. The effect of this decrease in water potential would be negligible, as can be seen from the following statement.

The relative lowering of the vapour pressure for ideal solutions obeys Raoult's Law, according to which

$$\frac{\text{lowering of vapour pressure}}{\text{vapour pressure of pure water}} = \frac{\text{no. of g molecules of solute}}{\text{no. of g molecules of solute} + \text{no. of g molecules of solvent}}$$

$$\text{For a solution of 1 g molecule per litre:} \quad \frac{1}{\frac{1000}{18} + 1} = \frac{18}{1018} = 0·0178$$

Therefore, the vapour pressure of a 1 g mol solution would be reduced by 1·78%. For instance, at 20 °C the saturated water vapour pressure is 23·3 mbar, the reduced value would be 22·9 mbar and, if the solution were one of 4 gramme molecules per litre, the reduction would be 6·7%, i.e. from 23·3 mbar to 21·8 mbar.

It must not be thought that all the salts entering the roots are transported in the transpiration stream to the leaves, and that only back-diffusion prevents their accumulation there. Most ions and other solutes are absorbed by various root, stem and leaf tissues, but especially by actively growing and developing cells.

Ethereal Oils. The rate of gaseous diffusion of molecules of a particular substance is inversely related to the density of the gas through which they are diffusing. Thus many arid-zone plants have evolved the production of ethereal oils, which slow down the loss of water by transpiration. These oils evaporate from the plant tissue in which they occur, especially the leaves, their evaporation rate increasing with temperature, so that the gas in the leaf air spaces and boundary layer has a higher density and therefore offers a higher diffusion resistance to water vapour from cell walls. Thus the water loss is reduced. Xerophytes from dry habitats, such as species of Eucalyptus, and xerophytes from physiologically dry (cold) habitats, such as conifers, are examples in which this mechanism operates. The many species of Citrus, which grow in warm dry regions but are not true xerophytes, are other examples in which numerous ethereal oil glands occur in the leaf tissues.

Calculation of transpiration rates

Using the expression

$$\text{rate} = \frac{\text{difference in density}}{\text{resistance}} = \frac{\mu g \, cm^{-3}}{s \, cm^{-1}} = \mu g \, s^{-1} \, cm^{-2}$$

$$\mu g \, s^{-1} \, cm^{-2}$$

the resistance at sunrise, including boundary layer resistance of $6 \, s \, cm^{-1}$, is assumed to be $15 \, s \, cm^{-1}$.

The resistance one hour later including a boundary layer resistance of $1 \, s \, cm^{-1}$ is assumed to be $2 \, s \, cm^{-1}$.

The density gradient at sunrise with the leaf and air temperature both at 16 °C will be $4 \, \mu g \, cm^{-3}$ if the air is at 70% of saturation.

One hour later, with the leaf at 23 °C and the air at 18 °C, and the percentage saturation decreased to 60%, the density difference will have increased to $11.15 \, \mu g \, cm^{-3}$.

Therefore the rate of transpiration per centimetre^2 of leaf surface will be:

$0.27 \, \mu g \, s^{-1}$ at sunrise and one hour later it will be $5.57 \, \mu g \, s^{-1}$
or $0.97 \, mg \, h^{-1}$ at sunrise and one hour later it will be $20.0 \, mg \, h^{-1}$

and expressed per dm^2

$0.097 \, g \, h^{-1}$ at sunrise and one hour later it will be $2.0 \, g \, h^{-1}$

For single leaves, or perhaps single plants with known leaf area, extrapolation of such calculations may give realistic values for transpiration rates. Extrapolation for crop canopies would, however, result in gross errors because of many modifying conditions found in a crop.

2.6. The measurement of transpiration rates

The measurement of transpiration rates of detached leaves or twigs and of potted plants is most reliably achieved by weighing, provided adequate care is taken to prevent or to assess separately any loss of water by evaporation from reservoirs or the soil.

In the case of detached parts of plants, an alternative to weighing consists of measuring the amount of water required to replenish the reservoir after the experimental interval. It must be remembered, however, that this can give average values only, over a period of time which should include an extra period during which, in the absence of transpirational vapour loss, no further water intake occurs, thus allowing for the "absorption lag" (p. 68). A suitable apparatus is shown in figure 2.5; it is essentially a modified potometer. This could be further modified for rooted plants, in which case the absorption lag will be more pronounced.

For rooted crop plants and plants in a field situation the apparatus just described has been developed into the *lysimeter* (figure 2.6). This is a hydraulic or mechanical balance sunk into the ground. On its "pan" there rests a representative soil volume with its vegetation, and this is continuously weighed. The balance pan carrying the load must be adequately drained so that it can be brought back to field capacity (p. 96) by the addition of water. The added amount of water represents the amount lost by transpiration and by evaporation from the soil over the experimental interval. Such information is valuable for irrigation

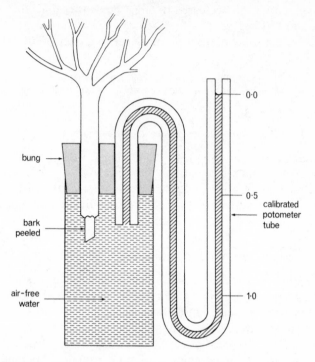

Figure 2.5. Apparatus suitable for the simultaneous measurement of rates of water vapour loss by weighing, and water uptake by potometer readings, using plant shoots or rooted plants.

work. A hydraulic lysimeter is shown diagrammatically in figure 2.6. Temperature corrections have to be applied and are troublesome.

A different principle is employed when transpiration rates are determined by measurements of water vapour lost from plant shoots. The method involves the setting up of somewhat artificial conditions in that an airstream of known moisture content is passed over the foliage in an enclosed system and the moisture gain determined. This can most conveniently be done by infra-red gas analysis or absorption in calcium chloride. A parallel stream without foliage is commonly used, and the transpiration rate calculated from the difference. For measurements of transpiration rate, evaporation from the soil should be prevented. The results are valid only for the conditions of air movement, atmospheric humidity, light and carbon-dioxide concentration in the air of the system employed during the measurements.

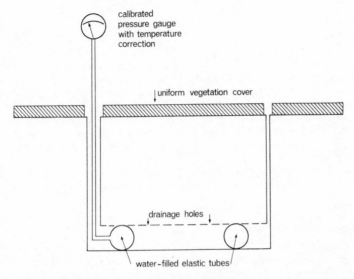

Figure 2.6. Hydraulic lysimeter. The temperature of the water in the system must be known for the temperature correction.

Other methods of measuring transpiration rates are based on infra-red analysis of boundary layers. An infra-red beam is passed through the boundary layer of a leaf and detected on the other side by a narrow infra-red photo-cell. Increases in the density and thickness of the boundary layer are recorded as decreases in the amount of infra-red radiation received at the detector. While this technique cannot easily be used to measure the absolute transpiration rate, it is very useful in measuring changes in transpiration, as reflected in changes in the boundary layer density, or changes in the boundary layer thickness as a result of a changing environment (e.g. the effect of an increase or decrease in wind speed).

An indirect method giving an estimate of transpiration rates of single plants consists of applying one of several methods designed to measure the rate of sap movement in the stem of a transpiring plant. Such methods are discussed in chapter 3. As a basis for transpiration-rate estimates, they suffer from the disadvantage of all "potometer-like" methods, which is that rates of water intake and rates of water vapour loss need not be the same, because of the absorption lag.

Experiments

Experiment E2.1. *Determination of leaf air-space volume.*

Strips of leaf tissue 0.5×2.0 cm are accurately weighed before being placed into a suction flask containing some boiled distilled water. Suction is now applied for about 3 minutes, and then suddenly released. Under suction air bubbles will be seen to escape from the tissue; when the suction is suddenly released, water will be driven into the air spaces. This is repeated several times until the appearance of the tissue indicates that it has been infiltrated. After gently drying the strips, they are again weighed accurately. Increases in weight represent the volume of the air spaces. The tissues are next dried at $105\,°C$ until a constant weight is obtained. From the data the air-space volume can be expressed as a percentage of the leaf water content or of its dry weight.

By using fully turgid leaves, and leaves with a known water deficit, changes in air-space volume with changes in leaf water content can be estimated.

Experiment E2.2. *The rate of evaporation as a function of the ratio:* $\dfrac{perimeter}{area}$ *and of speed of air movement.*

Several sets of shapes as shown in figure E2.1 are prepared from blotting paper and each moistened with $2\,cm^3$ of distilled water. The shapes are then suspended horizontally on suitable wire stands and weighed accurately every 5 minutes. One set is kept in still air and another in a gentle air stream. The data will show that the rate of vapour loss in still air is proportional to the perimeter rather than the area, whereas in moving air this distinction almost disappears.

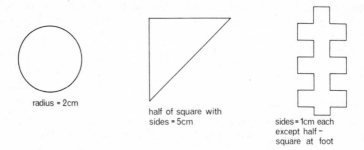

radius = 2cm

half of square with
sides = 5cm

sides = 1cm each
except half –
square at foot

Figure E2.1. Shapes of blotting paper of equal area ($12.5\,cm^2$) but different lengths of margin.

Experiment E2.3. *The cooling effect of transpiration.*

Rooted plants such as sunflower, tomato, broad bean, runner bean (dwarf), cockle-burr or maize are suitable. They can be made to transpire freely in artificial illumination and in a gentle breeze provided by a fan not directly aimed at the foliage. Thermocouples are used with the one junction in a constant-temperature flask (Thermos flask with water and ice) and the other addressed to the experimental leaf. Absolute leaf temperatures are thus measured, and comparisons can be made between the temperatures of leaves on a well-watered plant in direct light, and those of a well-watered plant in direct light which had previously been sprayed with a 10^{-8} M solution of ABA to keep the stomata closed. Leaf temperatures of well-watered transpiring plants and those of similar plants not watered for some time, so that the plant is under stress and with some leaves just about wilting, will be equally instructive.

Experiment E2.4. *Measurement of transpiration rate.*

The apparatus shown in figure 2.5 is suitable for these measurements. Either cut off shoots, or specially grown rooted plants can be used. In the latter case the stems must be held in split rubber bungs, which are air-tight. The apparatus is suitable for standing on most balance-pans, and losses in weight can be determined at intervals, either in still air or in moving air, in the light or in the dark, and in air at different water vapour content.

The movement of the meniscus in the capillary will give a measure of the rate of water intake by the plant.

Experiment E2.5. *Measurement of relation between diffusion resistance and path length.*

Three dry glass sample tubes of equal diameter, A, B and C, are filled with distilled water from a wash bottle: A to a height of 2 mm from the bottom, B two-thirds full, and C to within 2 mm of the rim. Immediately after filling, strips of cobalt chloride paper are placed across the mouths of the tubes. A large-enough beaker is inverted over the tubes, and the times measured for the papers to turn pink.

This simple experiment serves to make students think about diffusion path resistance.

CHAPTER THREE

WATER MOVEMENT THROUGH THE PLANT

IN A LIVING PLANT THERE ARE ALWAYS CONTINUOUS WATER COLUMNS in the xylem capillaries; in fact there is a continuity of water molecules from the soil outside the roots to the sites of evaporation in the leaves. As a plant grows, new xylem capillaries form at the ends of stems as hydraulic extensions of those already existing, and new capillaries are formed concentrically by the vascular cambium, causing the stems to increase in girth. These capillaries grow as living cells, the protoplasm of which breaks down after differentiation when they become water-filled conduits. Continuous hydraulic connections within a plant are also maintained during the growth and development of leaves and roots. Thus a plant can be thought of as a single, if complex, hydraulic system, all functioning parts of which always have a continuous liquid water phase between them.

Most of the water taken up by plants is lost as transpired vapour which evaporates from the wet inner epidermal and mesophyll cell walls in the leaf. If the plant is to remain viable, the lost water must be replaced so as to prevent death by desiccation, and in higher plants this replacement can occur only by water flow into, and through, the plant from the soil. This flow is caused by the lowered water potential in the leaves. In the previous chapter we considered the evaporation of water in the leaf, the path of water vapour out of the leaf, and to some extent the path of liquid water within the leaf. Here we consider how the lowered water potential in the leaf is transmitted through the plant, as well as some cases of non-equilibrium water flow through the plant.

3.1. Transpiration and plant water potential

The Path of Liquid Water in Leaves. Leaves of a small plant which has its roots in wet soil will have water potentials approaching zero if transpiration is not occurring, as the positive (root) pressure (p. 113) in the plant tissues will be opposite to the potentials (mainly osmotic) that cause root pressure. As soon as transpiration starts, the water potential at the evaporation sites in the leaf, principally the inner epidermal and mesophyll cell walls, will fall. Owing to viscous drag, there is a resistance to water flow through the plant, and therefore a potential difference will be set up between these sites and the water in the soil. The potential difference between the two ends of any part of the pathway will be proportional to the resistance to flow through it, and the sum of these individual potential differences will be equal to the potential difference between the leaves and the soil.

A fall in the water potential at the sites of evaporation in the leaf will cause a bulk flow of water through the wall immediately behind them, and ultimately into the mesophyll and epidermal tissues from the leaf veins. Flow within the tissues occurs principally in the apoplast, but though the resistance to flow through the symplast is about 50 times higher than that in the apoplast, there will nevertheless be some flow through this other pathway, as the two are in parallel. The proportion of water flowing through each pathway is not constant, but varies with the flow rate. Data have been collected indicating that the apparently non-linear hydraulic resistance in the leaf mesophyll is due to a decrease in water flow in the symplast pathway. This in turn is due to a decrease in cell expansion, and an increase in flow in the apoplast path when the transpiration rate increases and the leaf water potential falls. This accounts for most of the fall of approximately 30 times the measured hydraulic resistance of a whole plant from a low to a high rate of flow through it. *This* is true even though the length of the path through the mesophyll is fairly small in most species, as their evaporation surfaces are normally no more than two cells away from a xylem element.

In most species the large veins do not contribute appreciably to the direct water supply to the rest of the leaf. In fact the larger veins, though conspicuous, constitute only about 5% of the total vein length and are more important as structural elements and as "water mains" through which the smaller veins are supplied. The smaller veins ultimately ramify in single xylem elements, which in most leaves are so numerous that they constitute by far the most important route by which water is resupplied to the mesophyll and epidermal tissues. However, mesophyll cells near the larger veins obtain some of their supply through the bundle sheath extensions—sclerenchymatous tissue which surrounds the larger veins, but which occurs mainly between the veins and the epidermis.

All of the water lost from a leaf in cuticular transpiration, and some of that lost in stomatal transpiration, originates in the first instance from the epidermal cell walls. Thus the epidermis must be resupplied with water from the rest of the plant. In mesomorphic leaves there are two supply pathways directly from the vascular tissue. Water can move from the larger veins either through the bundle sheath extensions, which are much more important in resupplying the epidermis than in re-supplying the mesophyll, or through the very fine ultimate ramifications of the veins, the vein extensions, which form a network or fine mesh contacting the epidermis at points approximately 240–250 μm apart in mesomorphic leaves. In xeromorphic leaves the situation is very different as there are few or no bundle sheath extensions reaching the epidermis. The vascular bundles are embedded in the mesophyll, and the vein extensions form a much more widely separated network or are absent altogether. This means that in many xeromorphic species there must be considerable lateral water transport through the epidermal and mesophyll tissues from the relatively few points of contact with the veins. Alternatively, it is thought that isolated tissues may be able to satisfy their water requirements by absorption of water vapour from the leaf air spaces.

Water movement through the non-vascular tissues of leaves is both interesting and important to a full understanding of leaf water relations. In all leaves there is some paradermal flow through both the epidermal and palisade mesophyll tissues, the spongy mesophyll being too loosely packed for there to be significant flow in this direction through it. The resistance to paradermal flow in the mesophyll is lower than that in the epidermis, but as the latter tissue contains less water per unit area of leaf at a given degree of hydration, the rehydration "efficiency" of the epidermis is greater than that of the mesophyll if the tissue is re-hydrating solely by paradermal flow. In dorsi-ventral leaves there will be considerable flow in spongy mesophyll tissue at right angles to the epidermis. In mesomorphic leaves water flows from both the pallisade mesophyll and the epidermis to rehydrate the spongy tissue. However, in xeromorphic leaves water flows from the pallisade tissue to rehydrate the spongy tissue or to resupply the epidermis from the rest of the leaf.

The Path of Liquid Water in Stems and Roots. Water movement through the veins and the stele of the stem and root occurs mainly in the xylem capillaries, i.e. in the vessels and tracheids of the vascular tissue. This is a bulk flow phenomenon, as can be demonstrated by applying isotopically labelled water to the roots of a transpiring plant and observing that it rapidly replaces the water that was originally in the xylem. Water transport through stems is a purely physical process and does not involve living tissue. This was shown by Strasburger (1893) who killed the living cells of an excised stem with picric acid carried in the ascending water. Similarly, high temperatures (90 °C) do not prevent water transport in such a stem. Water transport occurs in the so-called *sapwood*, which is composed of the xylem elements of the outermost annual rings and associated parenchymatous tissues. In some species transport is almost entirely confined to the newest xylem ring. The heartwood in the centre of the stem is a supporting column with plugged xylem elements, which may partially decay.

Resistances to Liquid Flow. In stem, leaf, and root tissues of small plants, or in horizontal branches, pressure potential gradients (tensions) normally occur only as a result of transpiration, and are entirely due to the resistances to flow through the plant tissues. If such resistances did not exist, a fall in water potential at one point in the plant would cause water to move virtually instantaneously in response to the change in potential; the water potential would thus be equalized throughout the plant. If the resistance per unit length of water path in a plant, or part of a plant, is constant along its length, then the water potential gradient will be the same along any part of that pathway. If this is not the case, the greatest potential gradient and the greatest potential drop will occur where the resistance is highest. These gradients may be thought of as frictional or dynamic gradients.

In order to compare, or manipulate mathematically, water potential gradients in different parts of the transpiration stream, the resistances must be expressed in the same units in all parts of the path, as must the gradients (section 1.7). Experimentally determined diffusive resistances in the vapour phase are usually expressed as the time taken for diffusion to occur over unit distance (chapter 2), though as we have seen in section 1.7 they can be expressed as the potential difference between the ends of a path per unit rate of flow, i.e. $\dfrac{N\,m^{-2}}{m^3\,s^{-1}} = N\,s\,m^{-5}$.

These units are also used in determinations of the hydraulic resistance of xylem tissue, though the values obtained are usually expressed as a hydraulic conductance, which is the reciprocal of resistance (units $m^5\,s^{-1}\,N^{-1}$—the volume rate of flow per unit pressure difference between the ends of the tissue, see equation 3.1) and is a measure of the ease with which water flows through the tissue. A measure of the conductance of a piece of tissue may be useful when working with that piece of tissue, but the hydraulic conductances of two different pieces of tissue cannot always be compared, even if they are from the same plant, unless the tissue dimensions, or more particularly the pathway dimensions, are known. Therefore, when comparing the ease or difficulty with which water flows through various pieces of vascular tissue, new terms are defined which take into account the path dimensions, i.e. the *hydraulic conductivity* and the *hydraulic resistivity*.

The hydraulic conductivity ($m^4 s^{-1} N^{-1}$) is the product of vessel length and conductance per unit total lumen area (see equation 3.2), while the hydraulic resistivity ($N s m^{-4}$) is the reciprocal of this. The hydraulic conductivity of xylem can be determined by excising a length of stem or root tissue from a plant and forcing water through it under a constant known pressure. When the water is flowing at a constant rate, a radioactive tracer (usually $H_3{}^{32}PO_4$) is injected into the water supply at the high-pressure end, and the rate of movement of the radioactive wavefront recorded using G-M tubes positioned along the length of the tissue, with their windows near to it. Such readings can be quite rapidly repeated on any piece of tissue by washing out the $H_3{}^{32}PO_4$ with water and then injecting some more. The value for hydraulic conductivity is given directly if the distance between the G-M tubes and the applied pressure are known, as

$$\text{conductance} = \frac{\text{volume rate of flow}}{\text{pressure difference between ends of tissue}} \tag{3.1}$$

$$\text{conductivity} = \frac{\text{conductance} \times \text{vessel length}}{\text{lumen area}} \tag{3.2}$$

Then, if there is no lateral flow in the stem,

$$\text{velocity} = \frac{\text{volume rate of flow}}{\text{lumen area}} \tag{3.3}$$

From (3.1) and (3.3),

$$\text{conductance} = \frac{\text{lumen area} \times \text{velocity}}{\text{pressure difference between ends of tissue}} \tag{3.4}$$

Therefore, from (3.2) and (3.4)

$$\text{conductivity} = \frac{\text{velocity} \times \text{vessel length}}{\text{pressure difference between ends of tissue}} \tag{3.5}$$

Measured values of hydraulic conductivity in stems range from $0{\cdot}45 \times 10^{-9}$ to $4{\cdot}22 \times 10^{-9} m^4 s^{-1} N^{-1}$ in conifers; 1×10^{-9} to $35 \times$

$10^{-9}\,\text{m}^4\,\text{s}^{-1}\,\text{N}^{-1}$ in deciduous trees; and 40×10^{-9} to $70 \times 10^{-9}\,\text{m}^4\,\text{s}^{-1}\,\text{N}^{-1}$ in non-woody plants.

The hydraulic conductivity of xylem tissue can also be calculated from Poiseuille's equation if the radius of the individual xylem elements is known; but the values so obtained tend to be too large, as the wall pits of tracheids and end walls of vessels probably tend to control the flow of water through the xylem elements, and their walls are not perfectly smooth. [When determining the ease with which water will flow through stem tissue, some workers have calculated a parameter known as the *relative conductivity* (the quantity of water per unit area of stem passing through a 15-cm length of stem in 15 minutes under a pressure of $4000\,\text{N}\,\text{m}^{-2}$; the units of relative conductivity are m^2). This is not useful in the context of the present discussion, as it is not expressed in the units of $\text{m}^5\,\text{s}^{-1}\,\text{N}^{-1}$ or $\text{m}^4\,\text{s}^{-1}\,\text{N}^{-1}$, that we are using here for all parts of the plant pathway.]

Obviously the water pathways through the root cortex and leaf mesophyll are difficult to analyse directly in units of $\text{m}^4\,\text{s}^{-1}\,\text{N}^{-1}$. Values of hydraulic conductivity of leaf tissue are frequently given in units of $\text{m}^3\,\text{s}^{-1}\,\text{N}^{-1}$ (conductance per unit area); but these are not compatible with values derived for the xylem, as the path length is not included in the derivation, and the area term is frequently the surface area of the leaf, from which transpiration occurs, rather than the area of the hydraulic pathway through the leaf tissue. On the other hand, some values of the hydraulic conductance of whole roots have been obtained, e.g. young wheat plants have been found to have a root conductance of $2\text{–}8 \times 10^{-13}\,\text{m}^5\,\text{s}^{-1}\,\text{N}^{-1}$.

The lowest water potential in an actively transpiring plant is frequently found at the top, while the highest is found at the bottom of the plant; but this is not as simple as it might at first appear. The gradient often acts this way because the most actively transpiring leaves are at the top of the plant. These leaves receive more light and wind, and have a lower osmotic potential and a greater stomatal density than those lower down. Also, the total resistance to flow through the xylem is greater for leaves further from the roots. However, it is possible for one "side" of a plant to be at a lower water potential than the other, if the one side is receiving, say, a much higher radiation load. The development of a water potential gradient in this way is aided by the higher resistance to water movement across a stem than along it (pp. 62, 83), although there will be some lateral water movement in the stem and, especially while the gradient is being set up, there is frequently movement of water out of the "shaded" part of the plant into the other side.

Rates of Flow of Xylem Sap. The rates of linear water flow through conductive tissues vary considerably, depending on the species and the pressure gradients, from an undetectable rate to maximum values of approximately $45\,m\,h^{-1}$ in ring porous trees, $6\,m\,h^{-1}$ in diffuse porous trees and herbaceous plants, and 0.1 to $0.7\,m\,h^{-1}$ in conifers. However, the volume of water flowing through a particular piece of xylem tissue will depend on the area through which conduction occurs (mainly the xylem lumens) as well as on the velocity of flow. In gymnosperms, both the volumes and rates of flow will be relatively evenly distributed through a cross-section of active conductive tissue, as the wall pits offer the main resistance to water flow in tracheids. On the other hand, in xylem tissues which contain vessels with pores in their end walls, these end walls will have a much smaller degree of control over flow through the capillaries, especially as vessels are often fairly long, and it will generally be true that in tissues with intact water columns 7–8% of the conducting vessels, i.e. the largest, will carry approximately 60% of the total water flow.

Measurement of Sap Movement. The attempt to procure reliable readings of sap flow (velocity) or sap flux (volume) in whole plants has been of interest for some time, and has led to the introduction of several techniques.

The simplest of these, which was used in early experiments, involved the injection of dyes near the base of a stem, and the sectioning of the stem some time later to ascertain how far the dye had been carried in the transpiration stream. One major source of error in this technique was that the dye would be carried in the shock wave resulting from the breaking of the water columns when the stem was cut. More recently dyes have been replaced by radioactive tracers, and the velocity of the wavefront in the transpiration stream measured with radiation detectors. This technique is still used, especially in field experiments where, if a radioactive substance such as $H_3{}^{32}PO_4$ is used to ensure that it occupies only the free space, the simplicity of the technique largely outweighs any disadvantages. However, both of the above techniques suffer from the following disadvantages: that inserting a needle into a stem will break a number of the water columns, and that the injected substances may not easily reach all of the capillaries while, especially at low velocities, molecular diffusion of the injected indicator substance may in addition produce a significant error.

In the early 1930s Huber introduced a new technique. This used the rate of heat conduction in the transpiration stream rather than the rate of conduction of an injected substance. A heating coil and a temperature sensor were attached to a tree trunk with the heating coil a known distance "downstream" from the temperature sensor, and the trunk thoroughly lagged between the two. An electric current was passed through the heating coil for a short time to produce a "pulse" of heat. The time taken for the "pulse" to reach the temperature sensor was used to indicate the sap velocity. However, this obviously ignored the convective and conductive terms of heat flow, and it was found incorrect to assume that the velocity of the heat pulse through the xylem is identical with that of the sap. The technique was then modified so that the heating element was located between two temperature sensors, placed upstream and downstream from it.

The sap velocity could now be found in either of two ways. A heat pulse could be applied to the stem via the heating element, and the time of arrival of the pulse at the two detectors ascertained; or the heater could be used to heat the stem continuously and the equilibrium temperature of the two detectors measured. The sap velocity could be calculated from either of these two sets of data. The upstream temperature sensor was used to detect either the velocity of the heat pulse moving with the water, or the temperature rise due to heat carried in the transpiration stream, while the lower sensor indicated the velocity of the heat pulse, or the temperature rise due to the conduction and convection of heat moving against the transpiration stream.

These techniques have been used to indicate the sap velocity, but the sap flux can be estimated from these data only if the flow pattern in the stem is known. However, sap flux can be calculated directly using a variation of the above technique, if the amount of heat required to maintain a constant temperature difference between temperature sensors equidistant upstream and downstream from the heating element is measured.

The heat flow technique, with many variations, is used for the measurement of sap flow in stems ranging in size from large tree trunks to stems only a few millimetres in diameter.

The most recent technique for measuring sap movement depends on the principle (of magnetohydrodynamics) that a conducting fluid flowing through a magnetic field will have a voltage induced in it at right angles to both the direction of flow and the magnetic field, in much the same way as induction occurs in a solid conductor. The induced voltage is proportional to the magnetic field strength, the width of the channel or pipe through which the fluid is flowing, and the velocity of flow. If the shape and size of the pipe is known, the induced voltage can be considered to be directly proportional to the volume rate of flow. The conductivity of even distilled water is sufficiently high to take advantage of the above principle (as long as an alternating magnetic field, or non-polarizing electrodes, are used to prevent electrode polarization), so that if a magnetic field is applied across a stem with metal electrodes just penetrating the bark at right angles to the magnetic field, the voltage picked up by the electrodes will be proportional to the flow in the transpiration stream. The readings so obtained are proportional to sap flux, but the apparatus must be calibrated for a range of stem diameters, as its sensitivity is not linearly related to stem diameter in the way that it is to the diameter of a simple pipe.

3.2. Pressure potential gradients in the xylem

Static Potentials. If a pipe were filled with water, sealed at one end, and then stood upright with the sealed end uppermost and the open end just dipping below a free water surface, there would be a pressure gradient along the pipe, with the greatest pressure (ambient atmospheric) at the bottom. This gradient is due to the weight of the water in the column, i.e. it is caused by the action of gravity on the water column. The density of water varies somewhat with temperature and solute concentration, but a column of pure water, which has a density of approximately $998 \, kg \, m^{-3}$ at $20 \,°C$, acted on by gravity with an acceleration at sea-level of $9.78 \, m \, s^{-2}$, will contain a vertical pressure gradient of $9760 \, N \, m^{-2}$ or $0.0976 \, bar$ per metre. The gradient is usually approximated to $0.1 \, bar/m$ or $1.0 \, bar/10 \, m$, which is sufficiently accurate considering that the density of water may vary, as mentioned above, and the gravitational acceleration varies with both latitude and altitude. This is a static gradient and will not vary from one diameter of tube to another, or with any other parameter that affects the resistance to water flow, as long as the water column remains intact.

In the above example of a vertical pipe filled with water, the column could not be more than approximately $10 \, m$ high, as it is being supported by atmospheric pressure alone. However, if the open end of the pipe dipped below the surface of water in a sealed container, the water column could be any height, provided that the cohesive forces between the water molecules in the pipe and/or the adhesive forces between the water and pipe walls were not exceeded by the tension in the water column; and provided that the reduced pressure did not cause air to come out of solution and form bubbles. Either of these events would cause the water column to cavitate (see p. 64).

As water flow in the xylem elements is purely a physical process, and the long axes of the elements are often approximately parallel to that of the stem, the same argument that has been used above for a vertical pipe full of water can be used for entire trunks or stems that are growing vertically, i.e. there is a pressure gradient acting vertically along them. This is so despite the fact that tree trunks, especially as they become taller, do not always contain single continuous xylem elements from top to bottom. The water moves vertically from one vessel element to the next, mainly through the perforated end walls, and from one tracheid to the next through the pits in the cell walls, though water can, and does, move laterally across stems as well as along them, and in this case it moves through the pits in the walls of both the vessels and the tracheids and in the parenchyma rays. Thus in tall trees, the water potential at the top is considerably less than zero, even when transpiration is not occurring.

For example, the pressure potential in the xylem elements at the top of a tree which is about 100 m high (e.g. a redwood) will always be less than approximately -10 bar. Therefore, except under exceptional conditions, such as may occur when the atmosphere is saturated with water vapour which is condensing in, or on, the leaf and being absorbed by it, or when there is a positive root pressure (see p. 113), the total water potential of leaves in the crown of the tree cannot rise above approximately -10 bar. In practice, the water in the xylem and the leaf will not be pure, and their water potentials will be further lowered by the presence of solutes (chapter 1). A vertical pressure potential gradient of approximately 0.1 bar m^{-1} has been measured in trees using a pressure bomb (see p. 63). However, it has been suggested recently that at least this gradient, which exists under conditions of static equilibrium (steady-state conditions with no flow) may not be predominantly one of pressure, but rather of matric potential. This would mean that the pressure potential gradient throughout a vertical stem would not express itself and that there would be a matric gradient within the xylem.

It was suggested that the matric forces may be exerted by filamentous gel-like molecules, which are strongly hydrophilic, lining the xylem elements. If the amount of this substance increases with height, a vertical chemical potential gradient of approximately $0.1 \, bar \, m^{-1}$ could exist in the water in the xylem, rather than a pressure gradient. This suggestion is hypothetical, as it is not backed by experimental evidence. Whether the vertical gradient in water potential is due to a matric potential gradient, or to a pressure gradient, it is certainly due to the weight of the water column. It exists even when transpiration is not occurring, and when there is no water movement in response to the gradient.

Dynamic Potentials. The lowered water potential in a leaf resulting from transpiration steepens the water potential gradient from root to shoot, or causes a gradient to form in the majority of vegetation types which are too small to have a significant hydrostatic gradient in conditions of static equilibrium. This will cause water to flow upwards through the plant, and the rate of flow will be directly proportional to the dynamic water potential difference between the soil and leaves (i.e. total water potential difference minus static pressure potential) and it will be inversely proportional to the sum of the resistances in the water pathway.

As we have seen, resistances of stems are large enough to be measureable and to cause pressure (or tension) gradients to form along them. However, compared with resistances in other parts of the transpiration stream, those in the stem are comparatively small. The relative smallness of the stem resistances is illustrated by the fact that pressure gradients in the stem do not usually exceed $0.5 \, bar \, m^{-1}$, which is only five times the hydrostatic gradient in vertical stems, though gradients may reach $1 \, bar \, m^{-1}$ in some conifers, no doubt because of their higher hydraulic resistance.

In many species the greatest resistance to water flow through the vascular system occurs in the petiole, and this may be significant in helping to isolate one leaf from sudden changes in the water potential of others, or in tending to cause partial leaf dehydration and therefore, possibly, partial stomatal closure at high transpiration rates. The water potential existing in the xylem of a stem is normally about the same as that in the leaves immediately above it, though short-term changes in leaf water potential will not be fully reflected in the water potential of the xylem below the leaf, owing to the buffering capacity of water in the pathway (see p. 67). This means that xylem water potentials can drop to approximately $-150 \, bar$ in extreme conditions, and the pressure potentials will be little higher than this. As the pressure potential decreases, the water columns in the xylem will contract laterally slightly and cause a decrease in stem diameter. A diurnal change in stem or trunk diameter can, in fact, be measured in plants growing out of doors, with a minimum diameter in the afternoon, when water stress is greatest, and a maximum in the early morning (figure 3.1). This change in diameter is partly due to a shrinkage of the xylem, but a large proportion of the stem shrinkage, especially at relatively high water potentials (higher than $-15\,bar$) and in small stems containing a large proportion of living tissue, is due to a decrease in the volume of the living cells of the phloem and cortex and, in younger stems, the pith.

The relatively high moduli of elasticity of the lignified xylem cell walls and the water they contain causes them to change relatively little in size and shape at high water potentials, compared with the living tissues. But at low water potentials the latter become decreasingly important in their contribution to total stem shrinkage, because they probably shrink proportionately less as water potential decreases. The large tensions that are said to occur in the water columns in plant stems have been shown to be theoretically possible, and have been produced experimentally with specially prepared water columns. Thus the tensions that exist in plant stems do not overcome the cohesive or adhesive interactions between water and wall or between water and water (see pp. 4, 65).

So far we have not considered the conditions under which the pressure potential in a stem can become positive. This can occur during

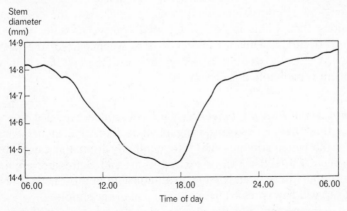

Figure 3.1. Diurnal changes in the stem diameter of *Gossypium arboreum* in clear weather.

the night, when the transpiration rate is extremely low (virtually zero) if the soil around the plant roots is wet. Water will move into the plant from the soil across the membranes of the root endodermis, and the pressure potential in the plant, that has been negative during the day, will become positive. This means that, given sufficient time, the pressure potential in the xylem can become approximately equal to the difference between the osmotic potentials of the xylem sap and the soil solution. When a positive pressure potential is developed in this way, it is generally referred to as a root pressure and, especially in smaller plants, leads to guttation (the extrusion of liquid water) from specialized structures in the leaves called hydathodes (see chapter 5).

3.3. Resistance to lateral flow in the stem

The resistance to water movement is of extreme importance in causing water potential gradients to develop and increase. The situation in stems is complicated by two factors: the path followed by the xylem elements, and the difference between the resistances along and across stems. Water flowing to a particular leaf always tends to move through a particular group of xylem elements, known as a *leaf trace*. These elements are different from those which supply other leaves, and usually occur in the stem immediately below the leaf they supply. However, the xylem elements frequently follow a spiral or other winding path, and the water flowing through their lumens follows the same route. At right angles to the long axis of the xylem elements, the resistance to water flow is approximately 50 times greater than the resistance to flow along them. This is not to say that transverse water movement does not occur in stems, as clearly it does when part of the pathway is blocked (e.g. by cavitation, see p. 64), but merely that transverse movement is probably not very great.

Measurement of Pressure Potentials. One consequence of negative pressure in the xylem of a transpiring plant is that, if a stem is cut or broken, the water columns will retreat rapidly from the exposed ends (Experiment E3.4). If a twig is excised in air and a pressure applied to its leafy end, the cut end being maintained at atmospheric pressure, xylem sap will be expressed from the cut end of the shoot.

This is the principle of the *pressure bomb* (figure 3.2), which is used to measure shoot water potentials by applying to the leafy end of a

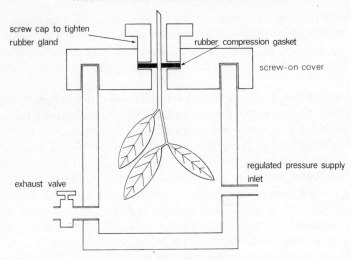

Figure 3.2. Pressure bomb.

cut shoot a pressure that is just large enough to force xylem sap along the shoot until it is level with the cut surface.

There are several, usually relatively small, errors involved in this technique, but it has some distinct advantages over other techniques. The advantages of the pressure bomb are that it provides a reasonably accurate estimate of shoot water potential very rapidly, and that an experienced operator can obtain a large number of repeatable readings from shoots of one, or many, species in a relatively short time. The apparatus is especially useful for field work, as it is robust, simple to set up and operate, can be operated from readily obtainable cylinders of compressed air or nitrogen, and does not require the closely controlled, or monitored, conditions needed for psychrometric determinations (section 1.7).

3.4. The transpiration pull

If the rate of evaporation from the cell wall surfaces in the leaf is suddenly increased, the resultant reduction in leaf water potential will cause water to flow through the rest of the plant to replace the water that is being lost. However, the absorption rate becomes equal to the rate of loss some time after the increase in transpiration rate. In addition, if the tension in the xylem becomes too great, the resistance to flow through it will increase. It has been observed that, even in a steady environment, the water relations of plants frequently do not reach a state of equilibrium, but change in a cyclical fashion due to changes in the plant. These matters are dealt with in some detail below.

The Ascent of Sap. So far we have discussed briefly, and in general terms, the forces acting on water moving in the transpiration stream, but have not considered how water can move through capillaries, especially the relatively large ones in the xylem, without the tensile forces causing the water columns to break. As pointed out in section 1.1, even though water molecules are in constant thermal motion with respect to one another, very strong intermolecular cohesive forces exist between them, and a tension of up to 350 bar can be transmitted through a water column confined to a tube without breaking the column. Similarly, there are strong adhesive forces between the water molecules and those of the lignified xylem cell walls, which prevent the "stretched" water columns from being pulled away from the walls. Thus the dynamic tensions caused by the removal of water from leaves by transpiration are transmitted through the xylem water columns to the roots, and cause a water movement through the plant. All water molecules that are ultimately transpired must take this route.

The preceding discussion of water movement through a plant in response to a reduced water potential in the leaves is a summary of the currently accepted theory of the ascent of sap. It is often called "The Transpiration Cohesion Tension Theory" (formerly Dixon and Joly's "Cohesion Theory of Ascent of Sap").

The production of tensions in the water columns in stem xylem has two important consequences. As the tension increases, the columns will become more and more "stretched" and will shrink laterally (p. 61) Also, conduction may be impaired by one of several factors, the principal one being cavitation.'

Cavitation. Thus far we have considered the transpiration stream as a simple physical system of water flowing through the root and the capillaries of the xylem to supply the evaporation surfaces in the leaf. As is very common with biological systems, water flow in the transpiration stream is not as simple as might at first appear, although apart from the root endodermis (p. 71), the transpiration stream can be considered as a physical system.

If the transpiration rate were increased, the water potential of the leaf tissues would fall, and this would generally tend to cause an increase in the rate of water flow through the plant, as well as an increase in the tension in the xylem. Although water columns can withstand high tensions, they will tend, for two reasons, to be disrupted as soon as significantly low pressure potentials develop within them. Firstly, the cohesive and adhesive forces will be considerably weakened, and secondly, as the pressure falls, there will be a tendency for air to come out of solution, or for it to be drawn into the vessels or tracheids through wall pits. It is unlikely that air will be drawn through pits until the pressure becomes very low (-300 bar or less) as the pits are very small. The pressure at which air does enter the vessels is known as the "air entry value". However, cavitation of water columns can occur. In fact, xylem water columns that are under tension are often "delicate", and a slight jarring of the tissue, for instance by wind, causes the columns to rupture. Thus it seems that the tension can exceed either the cohesive or adhesive forces, or that under such tensions air can come out of solution in the form of small bubbles. These would act as "flaws" in the column, weakening the cohesive forces. A vascular strand with a newly disrupted water column probably contains little air, but is filled with water vapour at a pressure determined by the water potential and temperature of the stem.

If the hydration of the stem tissues is much improved after cavitation, so that there is little or no water stress, at least some of the columns may reform. In fact, the resistance offered by the xylem to water flow can frequently be observed to fall when the stem tissue rehydrates. The reformed water columns are, however, often more fragile than others, as air in solution moves into the xylem elements when these are empty. Thus the reformed columns often contain small air bubbles, which do not much impede flow but which would act as "flaws" and cause cavitation to occur more readily if water stress recurred. At least the initial phase of any refilling process would be a physical one, probably due to the partial vacuum between the menisci "pulling" the water columns towards one another. However, it has been suggested that parenchyma rays may play some part in refilling cavitated vessels, as may inter-connecting bordered pits between a cavitated and a non-cavitated xylem element. Such pits may close during cavitation but reopen when the cavitated element starts to refill, thereby providing another path for rehydration, i.e. they may act as valves.

One way in which the conducting ability of plants is sometimes improved is by the production of more vascular tissue, which aids in the movement of water, or provides a substitute pathway to already cavitated tissue. In some ring-porous trees, capillaries in the xylem may be 15 metres or more long, and may extend almost the full length of the water pathway. Each newly produced vessel will differentiate in a few hours and will help to maintain a continuous supply of new water-conducting tissues.

When cavitation occurs, the surface tension of the menisci does not allow water to recede through the vessel end walls, or through the pits in the wall, even though the ensuing "shock wave" moves through much of the xylem tissue at a speed of about 0·4 to $1\,m\,s^{-1}$. This produces vibrations which, with the aid of sensitivity transducers, can be heard as "clicks".

Although the larger-diameter vascular elements are more liable to cavitation than smaller ones, it would be very unusual for all the water columns to be disrupted at the same time. Generally only a portion of the xylem elements in the sap wood are emptied, so that dry "islands" occur which are surrounded by xylem elements through which water is flowing. Although a measurable increase in resistance results from cavitation, it is as a rule not large enough to produce a great increase in water stress. This suggests that plants are "over-supplied" with vascular tissue.

This contention is supported by the results of experiments when overlapping cuts were made in the opposite sides of a tree trunk or branch. Provided the cuts were made further apart than a distance that is characteristic for each species, water could move around them (figure 3.3), and the water potential of the foliage was reduced little or not at all.

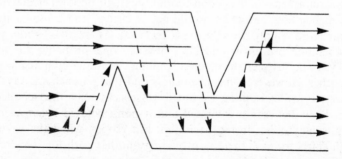

Figure 3.3. Water movement around overlapping cuts made in a branch. The arrows show the direction of water movement in the stem.

even though the resistance to water flow across a stem is much greater than the resistance to flow along it. If, however, the cuts were made closer together than the characteristic distance for the species, water flow was very much interfered with, and the water potential of the leaf canopy fell rapidly. In this case all, or most, of the xylem elements between the overlapping cuts were empty.

Cavitation is not the only result of water stress that can cause the resistance of the xylem to increase. Other increases in resistance are caused by deformations of the xylem elements under tension, and are directly reversible when the tension is removed. Vessel ends in leaves are relatively thin-walled; they will be the first part of the xylem to shrink, and may well collapse under increasing tensions, which will also cause closure of bordered pits in the xylem elements. Either of these effects will produce an increased hydraulic resistance in the xylem. The cessation of flow in one xylem element will not stop flow through other channels, because lateral transport can occur through vessel walls to allow a non-conducting capillary to be bypassed. The conductance along tracheid walls and parenchyma rays is in any case adequate to maintain the flow when only short lengths of xylem water columns are disrupted.

Time Lags in the Transmission of Changes in Rates of Water Flow. The diameter of stems changes with changing pressure potentials in the xylem. Leaf and root tissues also exhibit shrinkages during periods of water stress owing to transpirational vapour loss from the plant. In all parts of the plant there are two distinct reservoirs from which water can be lost.

1. The main path, which has a relatively low hydraulic resistance and which gives up and reabsorbs water rapidly.

2. The secondary path which has a higher hydraulic resistance and the hydration of which changes relatively slowly.

3.5. Hydraulic capacitance

We now discuss the amounts by which various tissues can be dehydrated and rehydrated in response to varying degrees of water stress, i.e. their *hydraulic capacitance*. This is the change in volume per unit applied pressure

$$\frac{m^3}{N\,m^{-2}} = m^5\,N^{-1}.$$

It must not be confused with the absolute amounts of water the tissues contain. These two values are very different, and their ratio varies in different parts of the plant.

The Absorption Lag. As well as having a "storage" or "hydraulic capacitance", from which water can be lost during periods of water stress, each part of the transpiration path also has a resistance to flow through it. The capacitance and resistance combined give rise to what is generally described as the *absorption lag.* If the transpiration rate of a plant suddenly increases, the rate of water uptake will not begin to increase simultaneously but will take some time before it equals the rate of water loss. The difference is made up from water supplied by the "capacitance" of the plant tissues.

Provided the values of capacitance and resistance do not change with changing water content of the plant tissues, a sudden cessation of transpiration will cause an approximately exponential rate of rehydration. If, in this case, the plant is considered as a simple system with a single capacitance rehydrating through a single resistance (figure 3.4), the time taken for complete rehydration would depend on the values of the resistance and capacitance, and on the amount of water that has been lost by the tissue. For an intact plant an approximation to a single capacitance and a single resistance is probably not unreasonable, as the roots are the site of by far the largest single resistance of any part of the plant through which water flows, and most of the capacitance occurs upstream of the roots. This can be demonstrated either by comparing measured values of root resistance with those of the rest of the plant or, more directly: the absorption lag in intact plants is much longer than in plants with their root systems removed. However, as we shall see, the plant cannot normally be considered to have a single capacitance and resistance.

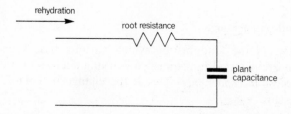

Figure 3.4. Plant represented as a single hydraulic capacitance rehydrating through a single hydraulic resistance.

In leaf mesophyll tissues, water in the transpiration stream flows mainly through the apoplast, which constitutes the main path, rather than through the symplast, which constitutes the secondary path. The degree and the relative magnitude of the dehydration of symplast and apoplast during water stress vary between species. Typically the apoplast will contribute about two thirds of the water for a fall in leaf water content of 10 to 20 per cent in a fully hydrated leaf, and this will result in a fall in leaf water potential of about 10 bar. Following a decrease in water potential at the evaporation surface, the symplast and apoplast will both become partially dehydrated, but the water content of the apoplast will re-equilibrate with the vascular tissue in 10 to 15 minutes, whereas the symplast will take approximately one to two hours (figure 3.5) owing to the higher resistance to water movement through the cell membranes out of the cells (p. 102). In this instance the symplast is best considered, not as a high-resistance pathway in parallel with the apoplast (p. 67), but as a capacitive element which is connected to the main pathway through a large resistance (the cell membranes). Re-equilibration with the vascular tissue when transpiration ceases takes similar lengths of time. Thus the cell-wall water is able to buffer the protoplasts from any sudden large changes in the water potential at the evaporation surface. In leaves, as in stems, there occurs a diurnal change in leaf water content, reflected in changes in both thickness and, to some extent, area, with a maximum at a little after midnight and a minimum in the mid-afternoon in clear sunny weather. This diurnal change is due to the relatively very high root resistance and reflects changes in whole-plant water content. However, changes in leaf volume occur very rapidly in response

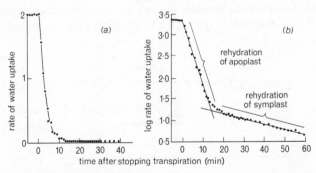

Figure 3.5. Water uptake into an *Ipomoea learii* leaf when transpiration is stopped by immersing the leaf blade in water: (*a*) absolute rate, (*b*) $\log_{10}$ of rate (see E3.5).

to changes in transpiration rate, and these volume changes are due to the combination of resistance and capacitance in the leaf. Leaf area may change by as much as 10%, and thickness by as much as 10–12% from maximum to minimum water content. When discussing the absorption lag, the pathway in the leaf mesophyll can be represented schematically in much the same way as that in the whole plant (see figure 3.4). However, in the case of the mesophyll the situation is more complex, as the mesophyll pathway cannot easily be reduced to two main components. If initially we consider only the apoplast, it is clear that this pathway has both resistance and capacitance, and for one cell wall from a xylem element to an evaporation surface this pathway may be represented as shown in figure 3.6. Adding the capacitance of the adjacent cell linked with this path through a large resistance, we have the situation illustrated in figure 3.7. For the pathway between a xylem element and the evaporation surface, there may be one or more of these elements in series, depending on the number of cells in this path. A large number of these paths are in parallel; however it should be noted that figure 3.7 represents the entire leaf mesophyll, provided that the values of resistance and capacitance in the diagram are the values per unit path length.

The transpirational flow in stems occurs almost exclusively in the xylem, which constitutes the main path for water movement, while the living tissues of rays, bark, and pith constitute the secondary path, which again is best considered as a capacitive element connected to the main pathway through a high resistance. Using the same considerations as when the leaf mesophyll was discussed, unit length of the xylem can also be represented schematically as in figure 3.7.

Relatively little is known about shrinkage of root tissues during periods of water stress, though it does occur. The main radial path through the root cortex is through the apoplast; the secondary path is

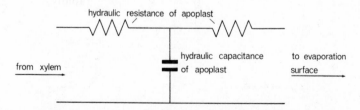

Figure 3.6. Schematic diagram of the transpiration stream within a leaf, between the xylem and the evaporation surface, taking into account only the apoplast pathway.

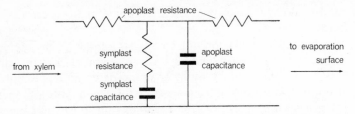

Figure 3.7. Schematic diagram of the transpiration stream within a leaf, between the xylem and the evaporation surface, taking into account the apoplast pathway and the symplast.

through the symplast; but in moving through the endodermis, water must pass almost exclusively through the symplast, as the apoplast is blocked by the Casparian strips. The part of the path through the endodermis contributes most of the hydraulic resistance of the root tissue and therefore of the plant. Once through the endodermis, water flows mainly through the apoplast of the pericycle to the stele, where the flow is largely longitudinal rather than radial. A schematic diagram of the root system is shown in figure 3.8, and differs from figure 3.7 only in having three resistance/capacitance elements, connected together through a high resistance. The endodermal resistance is very large, its capacitance is relatively small, and it does not have a secondary path component.

Many fruits, especially the more fleshy ones, for example citrus fruit, can show quite dramatic changes in size in response to changes in plant water-content. Although many such fruits have stomata and can transpire, in the context of this discussion they are again best considered

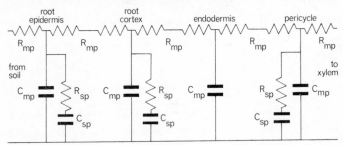

Figure 3.8. Schematic diagram of the transpiration stream within a primary root. R_{mp} represents the resistance to flow in the main path, C_{mp} the hydraulic capacitance of the main path, R_{sp} the resistance in the secondary path, and C_{sp} the hydraulic capacitance of the secondary path.

as an additional capacitance connected to the vascular system through a resistance.

The schematic diagrams for different parts of the plant hydraulic system are shown together in figure 3.9, which represents a plant with a single leaf, stem, and root. A more complex diagram representing a plant with four leaves, arising from a main trunk, and with a branched root system is shown in figure 3.10. This illustrates how some parts of the pathway are in parallel with one another, while others are in series; consequently mathematical analyses must be carried out with great care.

For example, if there is a steady flow through the "plant" shown in figure 3.10 and "branch" number one is excised, the flow rate through the "trunk" will immediately fall, while the proportion of the total flow through each of the remaining "leaves" will increase. At the same time, the potential drop along each leaf and individual stem will increase, and that along the trunk will decrease. If the flow rate is changing, the situation is even more complex. A sudden cessation of transpiration will result in a gradual fall in water uptake, as the tissue storage capacitances rehydrate through the resistances downstream from them. Similarly, a sudden increase in transpiration will produce a gradual rise in water uptake, as the tissue storage capacitances will lose water through the resistances on the upstream side of them. Thus the capacitance with the lowest resistance between it and the source of water or site of demand for water, will re-equilibrate fastest, so that those capacitances in the main path nearest the downstream end of the plant will reach a new equilibrium position first when transpiration falls, and those in the main path nearest the upstream end of the plant will re-equilibrate first when transpiration increases. If the rate of flow

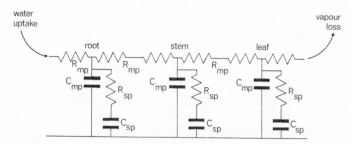

Figure 3.9. Schematic diagram of the transpiration stream in a plant with an unbranched root system and a single stem. The symbols are the same as those used in figure 3.8.

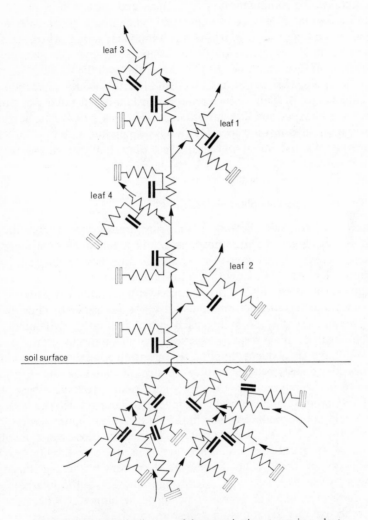

Figure 3.10. Schematic diagram of the transpiration stream in a plant with a branching stem, and a branched root system. The unfilled capacitor symbols represent the hydraulic capacitance of the secondary path, and the filled capacitor symbols that of the main path. For simplicity, the "return paths" for the electrical capacitor analogues are not shown.

of water through the plant is increasing and decreasing regularly (see next section), the combination of resistances and capacitances produces an impedance (a "resistance" to constantly changing flow which is of a greater magnitude than the resistance component alone) which causes both a lag in the transmission of the changes through the plant, and a reduction in their amplitude as they travel through the tissues.

The foregoing discussion has assumed that the values of resistance and capacitance of plant tissues remain constant when water potential, degree of hydration, and flow rate change. This, however, is not the case; both may change in quite complex ways, depending not only on the water potential and water content of the tissues, but also on the actual tissue involved.

3.6. Cyclic variations in plant water status

As we have said, water flowing in the transpiration stream encounters a number of resistances and impedances, and is regulated by a number of factors. These act as control mechanisms and may be external (soil water availability, light intensity, wind velocity, root and shoot temperature) or internal (diurnal rhythms regulating stomatal aperture, root resistance) and may interact antagonistically or synergistically. Thus, while most parameters may normally reach either a static or dynamic state of equilibrium, antagonism between two control mechanisms, or an overshoot and consequent "hunting" of one control factor, may produce a damped or sustained oscillatory change in water status. Further, because of the interactions between many plant processes, these will inevitably produce oscillations in a number of other parameters, although they would not normally oscillate in the absence of such interactions.

Probably the best-known types of oscillations are those which involve cycling of stomatal aperture. They are caused by instability in the feedback loop which controls guard-cell movements, either through changes in leaf water content, or through changes in the internal leaf CO_2 concentration. The former type of cyclic stomatal movement is also reflected in oscillations in transpiration rate, flow of the transpiration stream, plant water content, leaf temperature, and plant water potential. Such oscillations are characterized by periods of 20–40 minutes, although they can extend from 10–100 minutes or, in darkness, may even be as long as 2–$2\frac{1}{2}$ hours. Oscillations due to changes in CO_2 concentration have a period of approximately 2·5–4·5 minutes, and are thought to be directly related to guard-cell metabolism, so that while sometimes influencing the plant water balance they are not caused by changes in it.

Oscillations in stomatal aperture with a period of approximately 20–40 minutes or longer have been examined with reference to the mechanics of guard-cell movement, especially as affected by the water relations of the leaf epidermis. They can also provide information about plant hydraulics, especially the time lags that are involved, and the changes that occur in the values of hydraulic resistances and capacitances in the transpiration stream.

A less-well-known form of cyclic behaviour in plants involves cycles in transpiration and liquid water flow rates and in plant water-content, but without a significant concurrent change in stomatal aperture.

Stomatal Oscillations. If a plant is well watered, oscillations can be initiated by a sudden increase in transpiration rate, which will cause the leaves to become water-stressed but not sufficiently so for permanent stomatal closure to occur. Water stress will develop because changes in water uptake lag behind the changes in transpiration (p. 68). This may transiently produce a further stomatal opening, because the water content of the subsidiary cells adjusts to that of the rest of the leaf more rapidly than does that of the guard cells. Hence the ratio between turgor pressures of subsidiary and guard cells, which is one of the main determinants of stomatal opening, will be lowered. The larger stomatal apertures allow for further increases in the transpiration rate. Meanwhile, water uptake into, and flow through, the plant has begun to increase; but the rate of increase is too slow to prevent the leaf from becoming stressed to the point where loss of guard-cell turgor causes partial stomatal closure. Therefore, while the stomata are closing and thus lowering the transpiration rate, the rate of water movement into the leaf is rising (figure 3.11). The stomata may close almost completely during that part of the cycle in which their aperture is smallest, but frequently do not do so, the transpiration rate usually still being fairly high even at its lowest value, e.g. from a peak of $20\,\mu l\,dm^{-2}\,min^{-1}$ it may fall to only about $15\,\mu l\,dm^{-2}\,min^{-1}$. As might be expected, temperature cycles in anti-phase with the transpiration rate will develop.

Oscillations are not only initiated by sudden changes in the environment, but may occur spontaneously when external conditions are not changing. This occurs most often in the afternoon, and is thought to be due to an increase in root resistance at this time of day, causing the leaves to enter a condition of incipient drought, whereas earlier in

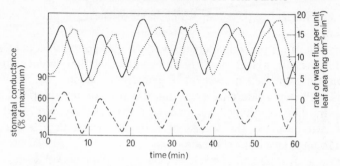

Figure 3.11. Oscillations with a period of approximately 10 min in the water balance of a leaf of *Gossypium*, showing cyclical changes in transpiration rate ————; water movement in the petiole ················; and stomatal aperture –––––––.

the day they had a water supply sufficient to keep them well hydrated even though transpiration might have been quite rapid. When oscillations occur spontaneously, they usually commence with a small amplitude which gradually increases.

Oscillations induced by illumination at the beginning of the day frequently die away, either because the transpiration rate becomes too high, or because the water supply is limiting, causing the leaves to become stressed and the stomata to close, more or less permanently. The oscillations may also become damped because the environment is too inconstant.

Oscillations may occur during the night, usually at temperatures above approximately 20 °C. The role of high temperatures in causing stomata to oscillate in the dark, and in enhancing oscillations in the light, is probably related to the higher respiration rate of guard cells at increased temperatures and to the dependence of stomatal opening on respiratory ATP.

Oscillations in stomatal aperture have most often been studied in small plants with all their leaves in the same environment. This has meant that the water balances and stomatal apertures of all leaves have oscillated in phase, partly because they all come under the same environmental influences, and partly because the water potential in the xylem oscillates and tends to force the oscillations into phase with it. This means that in a small plant, where distances in the hydraulic system are small, the time-lags in the transmission of changes in flow rate (p. 67) will be similarly small, and the changes in stomatal aperture, leaf water content, and transpiration rate in all leaves will be forced into phase by the oscillating xylem water potential.

In large plants the situation is more complicated because, although all the leaves of a plant will be influenced at the same time by large environmental changes, they will have different micro-environments and may experience different changes in these. Perhaps more importantly, there will be significant time-lags along and across stems between different leaves (p. 70). These lags increase with increasing periods of oscillation. Readings of oscillations in sap flux through a Morning Glory vine stem show a change in time-lag per unit length from $336 \cdot 7 \, \text{s m}^{-1}$ for oscillations with a period of 198 min, to a lag of $42 \cdot 2 \, \text{s m}^{-1}$ for oscillations with a period of 9·47 min. In fact, in large plants with a number of branches, the time-lags may be so great, and the micro-environments of leaves so different, that the stomata of leaves on different branches can oscillate with different periods and amplitudes. Such a situation often occurs when oscillations have been initiated by a sudden change in the environment (figure 3.12).

When the environment becomes constant and remains so for some time, the oscillations of stomatal opening and the water balance of leaves on different branches become more similar in period and tend to

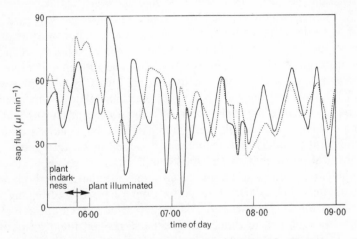

Figure 3.12. Oscillations in sap flux in two branches of a *Gossypium areysianum* plant. The branches were exposed to radiation intensities of $3 \cdot 39 \, \text{J cm}^{-2} \, \text{min}^{-1}$ ···················· and $2 \cdot 43 \, \text{J cm}^{-2} \, \text{min}^{-1}$: ————.
The distance along the branches and trunk between the points of measurement of sap flux was 64·1 cm. The angle between the branches, as seen when looking down the trunk, was 40°.

attain a relatively constant phase relationship. This is because the oscillating water potential in the xylem, acting as a "forcer", tends to unify the behaviour of different parts of the plant. The time-lags which occur in stem tissues result from both the hydraulic resistance and the capacitance of the tissues concerned. The fact that there is a lag in the transmission of changes in water content and flow rate, due to these components, might be expected to cause the amplitude of the oscillations to decrease down the plant. However, this does not occur, and the magnitude of oscillations in water uptake by roots is the same as that of vapour loss from the leaves. The reason for this is not clear, but it probably results from a change in the hydraulic resistance, and possibly the capacitance, of the tissues in the transpiration stream with changes in flow rate, or water content, or both.

Fluctuations in Leaf Water Balance. Fluctuations in leaf water balance are much less well documented than the oscillations discussed so far. The periodicity of changes in leaf water content varies rapidly (figure 3.13). often from one cycle to the next, and they are therefore better referred to as "fluctuations". They have been investigated mainly in single leaves mounted in potometers in leaf chambers with forced air-flow through them. In a few experiments intact plants have been used with their leaves exposed to the air, or with one leaf in a leaf chamber. In these cases the sap flux through the petioles was measured. Water vapour efflux from the leaves in chambers, or from leaves exposed to the atmosphere, has also been measured. The rate of water uptake, vapour efflux and leaf water content were all found to fluctuate with a period of 1–10 min. They did not show any correlation with changes in stomatal aperture, and only occurred if the transpiration rate and the water potential of the supply were high. A phase-lag occurred between the transpiration rate, leaf water content, and the rate of water movement into the leaf, such that water uptake and efflux were approximately 180° (half a period) out of phase with one another, and both were approximately 90° (quarter of a period) out of phase with changes in leaf water content. As discussed above, these time-lags can be explained on the basis of the resistance and capacitance in the transpiration stream in the leaf. However, to explain how and why sustained fluc- tuations occur in the first place is more difficult. They may result from a combination of the lags in the leaf and a positive feedback resulting from changes in resistance to water movement in the apoplast of leaf tissues, which is itself caused by changes in leaf water content.

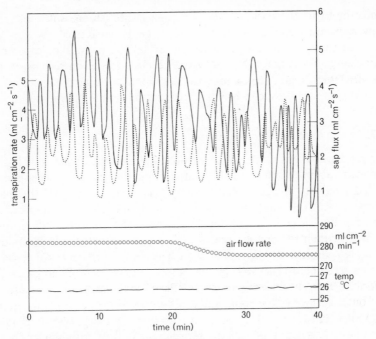

Figure 3.13. Fluctuations in water uptake into ───────── and water vapour loss from ·················· an *Erythrina indica* leaf fastened in a potometer while air was passed through the leaf under pressure. The viscous flow resistance of the leaf (stomatal aperture) ° ° ° ° ° °, and the air temperature ──────── did not vary significantly.

Between full hydration and a water saturation deficit of 5–7%, calculated for a fully hydrated leaf, the hydraulic conductance of leaf tissue increases. Thus a sudden rise in transpiration will produce both a fall in water content and an increase in hydraulic conductance. Hence, after a time-lag, rapid rehydration will occur. At the same time, the hydraulic conductance of the outer mesophyll cell walls will decrease, causing the transpiration rate to decline (see p. 43). As rehydration proceeds, the water content will rise and the hydraulic conductance of the apoplast will fall, slowing down rehydration; while the hydraulic conductance of the outer mesophyll cell walls will increase and the transpiration rate therefore increases. The inconstant period of the fluctuations is probably due to changes in water content, in water potential, or in the rate of water movement, since they affect the values of the hydraulic conductance and capacitance, and therefore the time-lags.

Thus the hydraulic conductance and capacitance change with changing flow rate and water content, so that if these two factors are not the same at the same part of consecutive cycles the periods of the cycles will change. Fluctuations are not likely to have deleterious effects on the plant, as they are not large enough to cause the leaves to become stressed at that part of the cycle in which water content is at a minimum.

3.7. Water movement across roots

As mentioned on p. 71, in the root most of the water must flow through the cell content of the endodermis, because flow through the endodermal cell walls is prevented by Casparian strips, and this path offers a relatively high resistance to water movement. Thus, during transpiration, a tension will develop in the xylem before water can be drawn through the endodermal cells of the root cortex sufficiently fast to match the rate of water loss, and a potential gradient will develop across these cells. The flow in the cortex occurs predominantly in the apoplast, though there is a much smaller flow through the symplast. At the outer surface of the root, water will move from the soil water films, provided that their water potential is higher than that of the water in the plant. The region of greatest water entry into the root is usually the root hair zone, 4–10 mm from the root tip. This region normally accounts for a large proportion of the water moving into small plants. However, as water stress increases, the region of greatest water entry moves up the root, away from the root hair zone, to parts that are suberized. Indeed, in large plants, especially trees, the suberized portions of the root system contribute a considerable part of the area of water entry (e.g. an estimated 93% in loblolly pine) and most of the water will pass through this portion of the root, even though the resistance per unit area is greater there than in the root hair zone. In fact it has been suggested that the function of the root hairs may not be so much to increase the surface area of the root to aid in absorption, but to maintain contact between the root tip and the soil when the latter's drying causes it to shrink away from the root. In view of this the figure quoted above for the proportion of water absorbed by suberized parts of the root is not so surprising as may at first appear.

When plants are actively transpiring and soil water becomes depleted, high rates of water use by the plant can cause the soil around the roots to become drier than the bulk of the soil. Under these conditions active root growth would not only increase the total absorbing area of the roots, but also enable the plants to tap new water supplies. The water potential of a plant growing in a soil which does not have a water input will decrease, gradually at first and then more rapidly as the soil dries out. Superimposed on this overall fall in plant water potential is a diurnal cycle, with the lowest potential in the early afternoon and the highest in the early morning. The soil water potential will also fall steadily at approximately the same rate as that of the plant, remaining equal to or slightly above the highest plant water potential of the daily cycle; however, it does not show a diurnal rhythm. As the soil water potential decreases, its hydraulic conductivity decreases even more rapidly (p. 90) so that a much larger difference in water potential is required between the soil and the root in order to move water into the plant.

The hydraulic resistance of roots varies. It falls by a factor of 2–3 times in response either to an increase in the steepness of the pressure potential gradient across the root, or to increased water flow through the root, or both, but there is also a diurnal change in root resistance, with its lowest value in the early morning and its highest in the late afternoon. It is also very sensitive to low temperatures. The inhibiting effects of low temperatures are important because the water moves through endodermal cells, whose permeability to water is much reduced below 5 °C in most species and below 10 °C in some species of warmer habitats. In fact, in regions where the soil remains cold without being frozen, while the sky is clear, this increased root resistance can be the sole cause of a plant suffering from a drought condition known as *physiological drought*, even though there is adequate water in the soil. Similarly, in arid regions where the temperature at night may drop to very low levels, the increased root resistance may slow down re-hydration of shallow-rooted plants, despite the upward movement of water in the soil due to the temperature gradient (see chapter 4).

We have shown that root resistance to water movement is influenced by potential gradients and temperature, but it is safe to assume that during periods when there is a positive pressure potential in the plant, root resistance to water movement is minimal. Root pressure has been thought to be essentially due to the osmotic movement of water through root cell membranes, especially those of the endodermis. However, there is evidence that root pressure is made up of two components, one of which is osmotic, while the other may be electro-osmotic.

Whatever the mechanisms are that cause a positive root pressure to be established in a plant, it can, as a rule, be measured only when there is negligible transpiration and when the water potential of the soil is higher than that of the plant. As transpiration increases, the rate of water inflow due to the difference in osmotic and matric potentials between the soil water and the plant sap is insufficient to meet the transpirational demand. This situation will cause the pressure potential in the plant to become negative. When a plant is transpiring, water flow into it is a function of the total difference in water potential between soil and plant water. The plant water potential is made up of an osmotic component and a pressure potential component. Since the pressure potential component is by far the largest and most variable when transpiration is occurring, it will usually be the largest driving force behind water entry into the plant. It is generally agreed that the osmotic component of water entry will contribute less than 5% to the total water movement in a rapidly transpiring plant. However, results have been obtained showing that the exudation rate from a detopped maize plant can be approximately equal to the transpiration rate of an intact plant.

3.8. Factors affecting water movement in plants

The major factors which determine the rate of liquid water movement within the plant are the rate of transpiration and the resistance to movement. As we have seen, these may be influenced by other factors, e.g. stomatal aperture (and therefore diffusive resistance) depends on light intensity and CO_2 concentration; root resistance depends on the temperature and degree of oxygenation; the water potential at the evaporating surface depends on the transpiration rate; while that at the root-soil interface depends on the water content of the soil. However, the two main factors interact, e.g. the stomatal aperture depends on the water content of the leaf, and therefore partially on the plant's hydraulic resistance. Root resistance depends on the water potential drop across the root and/or on the rate of water movement, as does the resistance of the leaf mesophyll, so that the water potential at any point in the plant depends on the resistance between that point and the bulk of water in the soil.

Experiments

Experiment E3.1. *Measurement of conductivity of a piece of plant stem.*

This experiment should be carried out with a leafless piece of stem tissue, freshly excised under water. The top end of the stem is removed and the lower end of the stem is dipped into a $0.1-0.2$ mM NaH_2PO_4 solution, which contains $H_3{}^{32}PO_4$ with a specific activity of $1-3$ mCi l^{-1}. The upper end is connected to a vacuum pump with a pressure gauge attached. The velocity of the radioactive wavefront is measured with a number of G-M tubes positioned along the stem. These readings can be made more accurate by fastening a collimator in front of each tube. The collimators are manufactured by making a slit $7-8 \times 1-2$ mm in a lead sheet. They should be positioned with their length at right-angles to the stem.

As the ^{32}P occupies the free space in the stem tissue, it can be washed out by dipping the lower end of the stem in water and applying the vacuum. By dipping the stem into the tracer solution once again, replicate readings can be obtained. The conductivity can be calculated from equation 3.5 if "vessel length" is replaced by "stem length" in the equation.

Experiment E3.2. *Demonstration of the heat-equilibrium technique of measuring xylem sap flow rates.*

The basic technique can easily be demonstrated in the laboratory with a small heating element, 12 mm in diameter, made of Nichrome or wire of similar resistance, and three thermocouples with identical resistances of approximately $3-8$ ohms. The heating element should be pressed firmly against a stem with a diameter of not more than about 3 mm. The three thermocouples should be positioned equidistantly (about 4 mm) above and below the element, to read the temperature of the heating element, and of the stem. The apparatus should be enclosed in a constant-temperature jacket at $20-25\,^\circ C$. A small current is passed through the heating coil to raise its temperature to about $14\,^\circ C$ above that of the air, and the currents produced by the thermocouples are measured with a small sensitive galvanometer after an initial equilibration time of $10-15$ min. The sap velocity is related to $\ln \dfrac{\text{upstream temperature}}{\text{downstream temperature}}$, so that both the absolute sap flow rate and changes in the rate can be recorded.

Experiment E3.3. *Lateral movement in stems under water stress.*

The experiment is best carried out with young *Coleus* plants. The root systems should be divided in two, and a portion of the part of the stem cut in half (figure E3.1). After the root systems have established themselves in their separate containers, one system is watered normally, while the other is allowed to dry out. During periods of good transpiration none of the leaves wilt, although some are supplied by vascular bundles directly connected to that root system which had been kept dry. (Vascular bundles and leaf-traces are easily traced in a cut split stem, one half of which dips into water containing methylene blue and the other into water containing Neutral Red dye.)

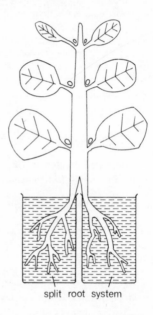

split root system

Figure E3.1. Apparatus used to demonstrate lateral movement in plant stems under water stress.

Experiment E3.4. *Demonstration of the existence of a tension in the stems of transpiring plants.*

If a branch of a rapidly transpiring plant is broken or cut under a dye solution, some of the solution will be taken up into the plant. If the branch is split along its length within a few seconds of being excised in dye, it will be found that dye has moved both up and down the branch and that, because little time has passed, the movement is almost entirely due to the release of tension in the xylem, and not to movement in the transpiration stream. This experiment can be carried out with most dyes and most species of plants, but works best with a low-molecular-weight dye such as 0·5% solution of methyl green, and with a woody plant, for example a *Eucalyptus* species.

Experiment E3.5. *Investigation of the uptake of water during the rehydration of the symplast and apoplast of leaf tissue.*

The apparatus required for this experiment is best constructed from a three-way tap connected to a fine capillary tube, to a larger tube which will be used as a reservoir, and to a tube to which the petiole of a leaf will be attached. The capillary tube is fastened horizontally next to a millimetre scale, and the apparatus positioned so that the reservoir stands vertically upwards, while the third tube hangs down (figure E3.2). A mature fully-expanded leaf, excised under water, is attached to the apparatus and allowed to transpire at a high rate for 1–2 hours. After this period a beaker filled with water at about 20 °C is raised from underneath

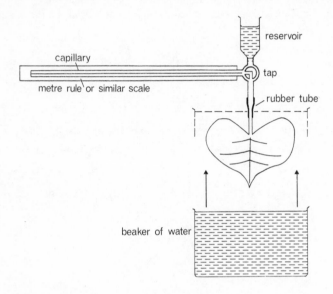

Figure E3.2. Apparatus used to show different rates of rehydration of the apoplast and symplast in a leaf.

to submerge the leaf entirely. The uptake of water from the potometer can be measured by recording the rate of movement of the meniscus in the capillary tube, and this will now be water which is rehydrating the tissue. The die-away curve obtained is logarithmic, and if it is plotted as the logarithm of the rate against time, a bi-phasal curve will be obtained which can be drawn as two straight lines. The more rapid and short-lived phase represents the rehydration of the apoplast, while the longer slower rehydration phase is due to water uptake into the symplast. (If the water in the beaker is at a temperature below 2 °C, the second uptake phase will not occur, as membrane permeability to water is very much reduced at low temperatures.)

These experiments are most easily carried out with large leaves, but to avoid anomalous results the low-temperature experiments are best conducted using leaves with a degree of frost resistance. The interpretation of results obtained in the cold-water treatment has been questioned. It has been claimed that the resumption of water uptake occurred on account of leaf expansion during the experiment. We therefore recommend the use of a "mature fully-expanded leaf". The possibility remains, however, that results reflect changes in leaf volume due to changes in temperature, and not changes in membrane permeability.

Experiment 3.6. *To investigate the absorption lag.*

This is most easily done with plants in whole-plant potometers. A whole-plant potometer can easily be made from a conical flask with a graduated side-arm and a tightly fitting rubber bung which has been split in halves and has a hole through it to take a plant stem. If an intact plant is put in the potometer

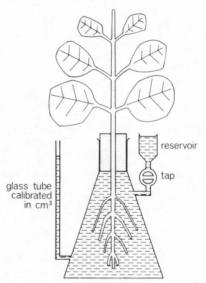

Figure E3.3. Apparatus used to show the large contribution of root resistance to the absorption lag. The plant stem passes through a hole in a split rubber bung.

(figure E3.3), the whole apparatus put on a balance and the plant allowed to transpire for 30 min or more, then covered with a plastic bag for 10–20 min, and then uncovered again, it will be found that the rate of water uptake (as indicated by the potometer reading) will not change as rapidly as the rate of water loss (as indicated by the change in weight). If the experiment is repeated using the same plant, but with the roots excised, the uptake lag will be found to be not so great.

CHAPTER FOUR

WATER IN SOILS

4.1. Soil water potential

The basic definition of the term *water potential* given on p. 10 should be kept in mind when applying it to soils. However, for water movement both within soils and into plant roots, only the difference in water potential between two points in the soil or between soil and plant root is of importance.

Soil water potential ψ_{soil} is a composite quantity because different forces contribute to lowering the potential of water in soils. There are adsorption forces between water and colloidal or mineral soil constituents, the latter especially when water films around particles reach multi-molecular dimensions. Added to these are surface tension forces in the water menisci of the soil pore spaces. Together these forces make up one component of the total soil water potential ψ_{soil}, namely, the matric potential ψ_m. Whether surface tension or adsorption predominates depends on soil type and texture, but also on soil water content. Generally, adsorption increases as soil water content decreases and, in dry soils, largely determines the matric potential by itself.

Another factor influencing the total soil water potential ψ_{soil} is the concentration of solutes in the soil solution. This ψ_π is thus a second component potential of the total soil water potential. In moist soils when the matric potential approaches zero, the osmotic potential may be the predominant component potential of the total. As a soil dries out, its matric potential becomes progressively more predominant in determining the total soil water potential, in spite of the fact that with drying the osmotic potential decreases significantly (p. 92). However, matric (adsorption) forces are capable of lowering water potentials more drastically than naturally occurring osmotic forces.

Another component of total soil water potential is a *pressure potential* ψ_p which is directly proportional to the excess hydrostatic pressure exerted over atmospheric pressure by the soil water column. This pressure potential can be of some importance at depth, but only in saturated soils. It must be noted that the pressure potential is positive and increases the total soil water potential. A somewhat related quantity, gravity, contributes a fourth component potential to the total:

$$\psi_g = \rho g h$$

From this expression it will be seen that the gravity potential is a function of water density ρ and the height above sea-level h, g being the acceleration due to gravity. For an elevation of 50 m with $\rho = 1000 \, \text{kg m}^{-3}$, $g = 9 \cdot 81 \, \text{m s}^{-2}$,

$$\psi_g = 1000 \times 50 \times 9 \cdot 81 \, \text{kg m}^{-1} \, \text{s}^{-2}$$

These are the dimensions of pressure (force per unit area) expressed in the units N m^{-2}, i.e. $\psi_g = 4 \cdot 90 \times 10^5 \, \text{N m}^{-2}$. Then, since 1 bar $= 10^5 \, \text{N m}^{-2}$

$$\psi_g = 4 \cdot 9 \, \text{bar}$$

which makes only a small contribution to the ψ_{soil} in a dry soil (cf. chapter 5).

Total soil water potential has thus four component potentials of varying importance depending on soil type, soil water content, and height above sea-level.

$$\psi_{\text{soil}} = \psi_m + \psi_\pi + \psi_p + \psi_g$$

Methods of measuring ψ_m and ψ_π are briefly mentioned in section 4.3; ψ_p and ψ_g must be calculated from basic information. For plant life, the total soil water potential alone is really of importance, and for practical purposes the relation between soil water content and total soil water potential is of interest because percent soil water content determinations can be made in most situations, whereas soil water potential determinations demand special equipment and take time (pp. 16, 97). The units of ψ_{soil} are bar where 1 bar is the pressure equivalent to a vertical column of water $10 \cdot 20$ m high.

If data such as are shown in figure 4.1 can be obtained for a soil under consideration, total soil water potential can be estimated and related to the water potential of plants growing in that soil at the time. For plants in contact with soil moisture, only the difference between ψ_{soil} and ψ_{plant} is vital. As long as ψ_{soil} is higher than ψ_{plant} water will flow into roots and continue to do so until the two potentials become equal. This will come about by gradual changes in pressure potentials when transpirational vapour loss slows down, i.e. negative pressures in

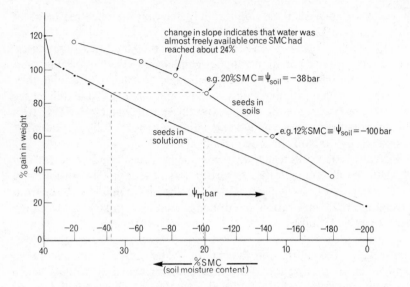

Figure 4.1. Percentage gains in weight of air-dry seeds kept in solutions of different osmotic potentials and of seeds kept in samples of soil of different moisture contents. By using this graph the soil moisture potential (or soil moisture tension) corresponding to any soil moisture content can be estimated for this particular soil.

the plant water system will change to positive pressures until they equal the difference between soil water potential and the osmotic potential of root cell sap. The following numerical example illustrates this.

Transpiring plant	Soil	Non-transpiring plant	
$\psi_{\pi sap} = -3\,\mathrm{bar}$		$\psi_{\pi sap} = -3\,\mathrm{bar}$	
$\psi_p = -7\,\mathrm{bar}$		$\psi_p = +1\,\mathrm{bar}$	← equal to the difference between ψ_{soil} and ψ_{sap}
$\psi_{plant} = -10\,\mathrm{bar}$	$\psi_{soil} = -2\,\mathrm{bar}$	$\psi_{plant} = -2\,\mathrm{bar}$	

$\psi_{plant} - \psi_{soil} = 8\,\mathrm{bar}$:
therefore net inflow of water into plant

$\psi_{plant} - \psi_{soil} = \text{zero}$:
therefore no net inflow of water into plant

4.2. Soil water movement

Soil science is a branch of science in its own right, and soil water relations a substantial part of it. We shall deal only with those aspects of water in soils that are immediately relevant to our treatment of water and plants.

Soils are heterogeneous systems of solid, liquid and gaseous phases, the latter perhaps commonly under-emphasized. Both inorganic and organic components of soils are represented in each phase, and they contribute to determining soil structure and texture. Any differences in structure between soils will greatly affect soil water relations. It is thus impossible to treat soil water relations quantitatively, so that what is said is valid for soils in general.

Leaving frozen soils aside, water may be present in the soil as water of crystallization, bound water, adsorbed water, water of hydration, and bulk water acting as solvent. Water in the soil originates both from precipitation, by infiltration and drainage, and by capillary rise from the permanent water table.

Infiltration. The initial movement of water into a soil is called *infiltration*; its driving forces are differences in matric potentials which decrease gradually, and gravitational potentials which gradually become more important. Infiltration hardly ever results in a saturated soil. The characteristics of the infiltration process and the rates of infiltration are of practical interest. Three zones can be distinguished: the saturated zone, the transmission zone, and the wetting zone with its wetting front. The water content in the saturation zone decreases rapidly with depth. In the transmission zone the water content decreases more gradually, and in the wetting zone and the wetting front it decreases once more rather steeply. For the wetting front to reach a depth of 10 cm may take two and a half hours in a certain soil; to reach a depth of 1·0 m may take 10 days, and to reach 1·5 m may take as long as 20 days. The process is thus rather slow. Besides gravity it is capillarity which drives infiltration or, as stated above, differences in matric potentials. In this context the term *hydraulic conductivity* is often used to describe the capillary properties of a soil. While this property is of great importance for infiltration rates, it is impossible to say anything generally valid about it, except that the hydraulic conductivity of a part-saturated or almost-saturated soil changes with changes in soil water content much more rapidly than does its water potential. For example, a change in the water content of a particular soil from 30% to 15% of dry weight is accompanied by a change in water potential from -0.28 bar to -12.0 bar, and a change in hydraulic conductivity from 10^{-6} to 5×10^{-9} cm s^{-1}. However, it must be remembered that each soil has its own characteristic value of hydraulic conductivity for a given soil water content.

In addition, the presence or otherwise of plant roots adds naturally to soil texture, and thus not only soil properties but also vegetation contribute to a soil's hydraulic conductivity. However, there is one factor which is generally valid and which greatly influences infiltration rates, irrespective of soil texture. This is the initial soil water content. As a rule, infiltration rates decrease as the initial soil water content increases. This can best be understood by remembering that rates of movement are proportional to differences in total soil water potential and, since this difference is greatest in a dry soil to which water has been added, e.g. during rain, the infiltration rate will be fastest in such a soil. The infiltration rate into clay soils is even more dependent on the initial soil water content than is the case with other soils. This is so because clay particles swell when wetted, thereby reducing the volume of the soil pore spaces between them. A wet clay soil will have less pore space into which water can penetrate than a dry one.

Retention. Water retention in moderately moist soils is primarily due to surface tension forces which develop at liquid/air interfaces when air enters the pore space, as water is lost from the soil by evaporation or movement into plant roots. Osmotic forces, which are important when soils become very dry and which are enhanced in soils which shrink on drying, are referred to below. For moderately moist soils the use of the theory of capillarity (pp. 4, 28) enables one to estimate pore size space in a drying soil by measuring the increase in suction with a Tensiometer (pp. 97, 99).

Taking as reference point pure free water at atmospheric pressure, the suction $(-P)$ at the measuring-point will be given by

$$-P = \underset{\substack{\text{water}\\\text{density}}}{\rho} \times \underset{\text{gravity}}{g} \times \underset{\text{depth}}{h} = \underset{\substack{\text{matric}\\\text{potential}}}{\psi_m}$$

$$\underset{\substack{\text{dimensions of}\\\text{force per unit}\\\text{area} = \text{pressure}}}{kg\,m^{-3} \times m\,s^{-2} \times m \qquad kg\,m^{-1}\,s^{-2}}$$

Alternatively ψ_m can be related to the radius of curvature of the capillary pores (assumed cylindrical as a first approximation, $2\pi r\,\text{S.T.} = \pi r^2 . \rho g h$). Using the relationship above we can state

$$\psi_m = -P = 2\,\text{S.T.}r^{-1} = \rho g h \quad (\text{S.T.} = \text{surface tension of water})$$

and from this

$$r = 2\,\text{S.T.}\,(\rho gh)^{-1} \text{ (see pp. 4 and 28).}$$

Thus we can estimate the maximum radius of soil pore space which can remain filled with water at the measured matric potential. Such calculations demonstrate by how much plant water potentials have to decrease for water to continue to flow into roots.

If such studies are carried out, and estimates are made of the various parameters involved, discrepancies will be observed in the results obtained, depending on whether the measurements are made in soils progressively drying or becoming progressively wetter. These discrepancies are hysteresis effects thought to occur because of irregularities in emptying and filling of the intricate network of soil pore spaces. Generally it is found that the measured water content is higher in a drying soil at the same ψ_m than it is in a wetting soil (figure 4.2).

As indicated above, an osmotic effect comes into play as soils dry out severely. The causes for such increased osmotic effects are complex and are more pronounced in soils which shrink on drying than in

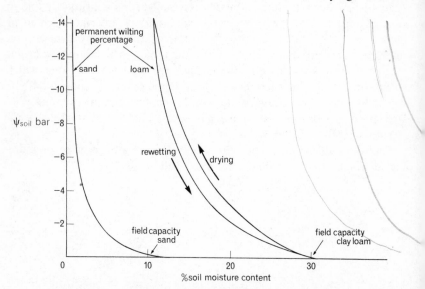

Figure 4.2. The relation between percentage moisture content of a soil and its water potential. The data apply to sand and a clay-rich soil. In the latter the measurements were made as the soil dried out gradually and when it was progressively wetted. The curve is known as a *hysteresis loop*. There is no hysteresis loop in sand.

others. During shrinkage, charged micellae, their adsorbed ions and double layers of hydration are brought closer together. As a consequence, the available spaces between micellae come to contain highly concentrated solutions with vastly decreased osmotic potentials, so low in fact that they significantly depress the vapour pressure of the soil solution (pp. 43, 87) with serious consequences for water availability to plants.

Movement of water in soils is continuous following differences in ψ_{soil}, with the component ψ_g more important in saturated and ψ_m in unsaturated soils. The depth of the water table influences movement to some extent. At equilibrium and in the absence of water removal by evaporation or by roots $\psi_m = \rho g h$, i.e. at every point in the soil profile ψ_m corresponds to the distance from the water table.

Drainage. Besides studies of the infiltration process and its rate, studies have been made of water movement in saturated and unsaturated soils. In both these cases Darcy's basic concept is applicable: the quantity of water passing a unit cross-section of soil in unit time is proportional to the difference in soil water potential between two points in the soil. This statement is not unlike that characterizing diffusive flow (p. 33):

$$\frac{\text{Rate of}}{\text{movement}} = \frac{\text{Difference in soil water potential} \times K}{\text{Difference in gravitational potential}} = \frac{\Delta\psi_{soil} \cdot K}{\Delta\psi_g}$$

In this equation ψ_{soil} is analogous to $\Delta\rho$, $\Delta\psi_g$ to the length of path (resistance) and K to **D**. The difficulty with such an expression is K the hydraulic conductivity (p. 90). Unlike **D** it is not a temperature-dependent constant, but a variable term depending both on the soil type and on the state of water in the soil. Hence the real task is to evaluate K. Procedures to estimate K are complex and beyond the scope of this text. Generally, in moist soils when the total soil moisture potential is high, K also has a relatively large value, and in a drying soil with very low soil water potentials (-10^7 bar) K is also a very small number indeed (10^{-11} cm s^{-1}) (p. 90). As stated in the previous chapter, total soil moisture potential is a composite term. As soils dry out, gravitational potentials become less important and matric potentials more so.

Capillary Rise. Considerable interest attaches to water movement upwards from the water table against gravity. Such capillary rise does occur, and its rate has been estimated for different soil types. Naturally capillary pore size is a determining factor, but the rate depends on the distance from the water table and the total soil water potential. Thus

Slatyer (1967) quotes Wind who estimated capillary rise to vary in rate between 1 mm and 5 mm per day. Once the total soil water potential rises to -1.0 bar in a fine soil and -0.1 bar in a coarse soil, rates of capillary rise become independent of soil water potential.

Water Vapour in Soil Pore Spaces. We have so far considered liquid water in soils. However, water vapour will be present in all pore spaces not filled with water, as long as there is some water in the soil. This vapour is important to plants in several ways. Once a pore space is at saturation water vapour density, roots will not lose moisture by evaporation. The sensitivity of saturation water vapour density to temperature makes the soil water vapour content a very dynamic parameter, with evaporation and condensation occurring constantly.

The fluctuations in soil temperature depend on the radiant energy input from the sun, on changes in air temperature, and also very much on soil moisture content. In moist temperate climates diurnal changes in soil temperature are generally restricted to the upper 5 cm of soil, but in arid environments such changes may be quite great at depths in excess of 20 cm.

At the soil surface in clear weather the maximum temperature occurs in the early afternoon and the minimum at sunrise. At increasing depths these "peaks" and "troughs" occur later, and have a diminished amplitude. These temperature fluctuations have a negligible effect on liquid water movement, which is normally upwards because of evaporation, but they can cause a net movement by "distillation". During the hottest part of the day, the water vapour density will be higher in the surface soil layers than in those below them, and this will cause a net downward vapour flux. As the lower soil layers are cooler, the water vapour will condense. The loss of water vapour to the atmosphere will reduce this effect. During the night the soil surface temperature will drop below that of the soil below it, thus reversing the effect. This phenomenon may be of considerable importance for supplying water to shallow-rooted plants at night in arid regions.

Evaporation from a Soil Surface. Evaporative loss from soil surfaces depends on the moisture content at the soil surface, because this will determine the water potential of the source, which is mainly a function of the matric potential. If the surface is wet, the evaporation rate from the soil will not be too dissimilar from that at a free water surface, since the water potential of a pool or film of soil water will be lowered only very slightly by a small osmotic potential. This situation will be maintained until the surface films of water begin to retreat into the soil, provided that the rate of evaporation does not exceed the rate with which water can move from lower soil levels to the surface films. If this proviso is not fulfilled, and the soil surface becomes gradually drier, secondary effects intervene which increase the rate of evaporation. The most important of these is the thermal nature of the surface; the thermal properties of the soil when wet are similar to those of free water, with much of the incoming energy absorbed in latent heat of vaporization and the high specific heat of water. However, when the soil becomes dry, its temperature rises much more steeply. In consequence, the water vapour density difference between source and atmosphere steepens, with consequent increases in the evaporation rate.

The effect of wind on the rate of evaporation from a soil surface is also similar to that from a free water surface; but when the soil surface is not wet, the relationship between wind speed and evaporation rate becomes more complex. In strong winds, which produce turbulence close to the soil surface, the resistance to water vapour diffusion within the top layers of soil decreases, because the turbulence produces pressure fluctuations in these top layers, so introducing some bulk vapour flow in addition to diffusive flow. On the other hand, increased streamlined airflow over a dry soil surface has relatively little effect on the evaporation rate, because the main resistances to loss occur in the dry surface layers.

4.3. Methods of determining soil water status

Reference to more specialist texts will show the existence of a very extensive, not to say bewildering, variety of soil water terms, some being used more consistently by soil physicists, others by engineers, agriculturists or other specialists. In this book we restrict ourselves to a small selection of terms which are relevant to plants.

Availability. In spite of the fact that for plant life importance attaches only to that fraction of total soil water that is available to plants, the division into *available* and *unavailable* water is unsatisfactory. It is true, water of crystallization and some of the bound and adsorbed water is rarely, if ever, available to plant roots, but the proportion of the total soil water that is unavailable depends in the first place on the relation between soil water potential and plant water potential. Both are changing quantities. As long as the soil water potential is higher than the plant water potential, some water will be available to the plant. However, it is not equally available at different values of ψ_{soil} and ψ_{plant} because it is the difference between these two values which represents the driving force for water to move into the plant. It is incorrect to assume that between field capacity and permanent wilting percentage (see below) changes in soil water content are unimportant and have no bearing on the availability of water to plants.

Percent Soil Moisture Content. This measure is probably the most directly available and practicable under most situations. It does not, however, indicate the force required for the removal of water from the soil into the plant. Percent soil water content is determined gravimetrically by drying a representative sample to constant weight at 105 °C. Soil water content is expressed as a percentage calculated on dry weight. The relation between soil water content and soil water potential referred to above is shown for a particular soil in figure 4.1 (cf. Expts. E4.1, 4.3).

Field Capacity. A special case of percent soil moisture content is known as the *field capacity*. It is an important soil characteristic greatly influenced by soil constituents and texture. Field capacity is the amount of water retained in a soil after drainage towards the water table has resulted in a near-constant soil moisture content. Although for all practical purposes the soil moisture content at field capacity remains constant, some water movement continues to occur at field capacity, although the rate is very low. A standard procedure to measure field capacity is to flood a representative soil sample and allow to drain for three days in the absence of evaporation before gravimetric determination of water content (Expt. E4.2).

At field capacity, total soil water potential will be high because with most pore spaces filled with water both osmotic and matric potential will be high as well.

Permanent Wilting Percentage. This concept has often been used as a soil constant; it should be accepted as such only if it has been determined under standard conditions with a standard test plant. The procedure employed must be rigorously defined if the value determined is to have any interest for comparisons between soils. The permanent wilting percentage is the percentage of water (calculated on dry weight) remaining in the soil after a specified test plant has wilted under clearly defined conditions of temperature, light intensity, atmospheric humidity and air movements, so that it does not recover, even if moved into a dark and cool place, unless water is added to the soil. Soil moisture content is again measured gravimetrically.

Using water potential terminology, the permanent wilting percentage is reached when the total soil water potential is equal to, or lower than, the plant water potential, i.e. the plant cannot gain any water from the soil unless the soil water potential is raised. The water potential of a wilted plant depends almost entirely on its osmotic potential (in which we include the cytoplasmic matric potential) because there are no positive pressure potentials when the plant is wilted. The osmotic potential of the plant is unlikely to decrease further after wilting. If it could, this would provide a difference between soil water potential and plant water potential, permitting some water movement into the plant.

Total Soil Moisture Stress. This concept is closely related to the method employed to determine it. TSMS is more closely related to soil water potential than the former three parameters. It is determined by suction plate, pressure membrane or Tensiometer. TSMS is given directly in bars when using these instruments. Other methods for its determination must be calibrated against any of these direct ones. Such methods include thermal conductivity probes, porous electrical conductivity blocks, and the neutron probe.

Soil Water Potential. This basic and all-important concept has been dealt with in section 4.1, and in this section we describe its measurement only; it should be noted, however, that soil water potential is a good measure of TSMS provided the soil is unsaturated.

The most reliable method for determining total soil water potential is to determine the water vapour density above a representative soil sample kept in a closed container at absolutely constant temperature. The measurement can only be carried out after a suitable equilibration time. The instrument most commonly used is Spanner's Peltier Effect Psychrometer. The instrument must be calibrated against known osmotic solutions with known depressions of water vapour densities (p. 15).

Two component potentials of total soil water potential, namely the matric and osmotic potentials of soils, can be determined separately. The remaining two, the gravitational and the pressure potential, must be estimated by calculation. The matric potential of a soil is in fact measured with suction plate or pressure membrane. The osmotic potential of the soil solution can only be measured by preparing a soil extract, in itself a risky and doubtful procedure. However, if this can be done satisfactorily, Spanner's Peltier Effect Psychrometer or cryoscopy are the most suitable methods for the determination of osmotic potentials (Expt. E4.3).

Possibly the most easily accessible method of determining soil water potential is a bio-assay method using the seeds to be planted in the soil under investigation. The procedure is simple. The seeds have to be calibrated by immersion in physiologically inactive osmotic solutions, and their gain in weight determined when an equilibrium has been approached, usually after about 12 hours. At the same time another batch of seeds is placed in a series of soil samples of graded moisture contents, for a period which does not allow germination, and gain in weight of these seeds is determined. A graph of changes in weight against % soil water content and ψ_π provides the desired information (cf. figure 4.1 and Expt. E4.3).

Experiments

Experiment E4.1. *Determination of percentage soil moisture.*
A representative but small sample of soil (3–5 g) without unusually large fragments of stone is weighed in a tared crucible with lid. Crucible, lid and sample are then dried at 105 °C till of constant weight. The use of desiccators is necessary. Results are expressed as percentage moisture content calculated on dry weight, i.e.

$$\frac{\text{loss in weight} \times 100}{\text{dry weight}}$$

Experiment E4.2. *Determination of field capacity of a soil.*
A representative sample of soil (5–10 g) is weighed into a tared Gooch crucible fitted with a piece of moist filter paper covering the perforations. The filled crucible is placed in a clay-triangle on a tripod and filled to the brim with water. This is repeated a few times until the water drops from the perforations. The crucible lid is kept in position to prevent evaporation from the top layer during the 72 hours during which drainage is allowed. Percentage soil moisture content is then determined as above.

There are several variations of this procedure. When centrifuging is used (at standard speeds for standard times) similar results are obtained; they are referred to as *soil moisture equivalent* not field capacity.

Experiment E4.3. *Bio-assay of soil water potentials.*
Suitable seeds are cocklebur, pea, runner bean, sunflower. Two lots of eight

batches of ten seeds each are weighed. One batch each is placed in water, 0·5 M, 1·5 M, 2·5 M, 3·5 M, 4·5 M and saturated solutions of a salt, or a series of polyethylene glycol solutions at about 15 or 20 °C. The other eight batches of ten seeds each are placed into samples of the same soil but of different soil moisture content. The soils are best kept in plastic bags and the seeds distributed so that they do not show. Percentages of soil moisture to be aimed at should be evenly spaced, e.g. 4, 8, 12, 16, 20, 24, 28 and 32%. Exact determinations have to be carried out after the experiment. The seeds kept in the solutions are removed after about 12 hours, dried and weighed; increases in weight are expressed as percentages calculated on starting weight. The seeds kept in the soil samples are removed after 48–72 hours; they must not be allowed to germinate properly and to produce primary roots. Percentage gains in weight are plotted against osmotic potentials of the solutions, and the resulting graph will resemble figure 4.1; from it the soil water potential appropriate to each percentage soil moisture content can be read off.

Experiment E4.4. *Measurement of soil water potentials with gypsum blocks and Tensiometers*

These devices can be usefully combined for the measurement of soil water potentials. Tensiometers will only work satisfactorily in the range of approximately 0 to −0·8 bar, while gypsum blocks are most sensitive over a range of about −0·5 to −15 bar. The gypsum blocks can be used at lower water potentials, though with a decreased accuracy.

Gypsum blocks are small square or rectangular pieces of gypsum 2–3 cm wide, 3–6 cm long and 1–2 cm deep, in which two electrodes have been embedded. The resistance of the material between the electrodes changes with changes in its water content, and these changes are best measured with an a.c. bridge. Gypsum blocks can be cheaply obtained ready made, but should be individually calibrated. This is best done by placing the blocks in the type of soil in which they are to be used, wetting the soil, and reading the a.c. resistance of each block as the soil is allowed to dry slowly. At the same times as the resistances are read, soil samples should be taken from very near the blocks and placed in a thermocouple psychrometer to determine their water potential. If a thermocouple psychrometer is not available, the water content of the soil samples could be obtained by weighing and drying, and the blocks calibrated for soil water content. Both during calibration and when in use in experimental situations, the blocks should be allowed approximately 24 h initial equilibration time, and they should be re-calibrated frequently, especially when used in acid soil.

A Tensiometer can be constructed from a ceramic cup, with its open end fastened (with epoxy or similar cement) into one end of a plastic tube. To the other end of the tube should be fastened a means of filling the pot with water, and a means of measuring pressure changes, such as a mercury-filled U-tube mano-meter, a vacuum gauge, or a pressure transducer. The apparatus is used by completely filling it with boiled water, closing the tap at the top, and inserting the ceramic cup in the soil. However, it cannot be used to measure potentials below approximately −0·8 bar as air bubbles form in the system, breaking the water column (figure E4.1).

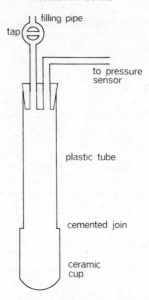

Figure E4.1. Soil tensiometer.

These two instruments together are excellent for investigating how rapidly a plant in a sealed pot dries the soil, or for following changes in soil water content while monitoring changes in plant behaviour. Their weakness is that both require equilibration, so that results obtained with them do not represent the state of soil moisture at the time of measuring, and the gypsum blocks exhibit both a degree of hysteresis and a change in calibration after a number of wetting and drying cycles.

Although we have referred to the parameter measured by these two devices as soil water potential, it is in fact a measure of all the component potentials of soil water potential except the osmotic potential.

CHAPTER FIVE

WATER IN CELLS AND TISSUES

5.1. Water in cell walls and intercellular spaces

The water component of single cells and tissues, although continuous, can be considered as distributed over two metabolically and physically distinct regions: (1) water contained within the protoplast and totally enclosed by cytoplasmic membranes; (2) water in contact with these membranes, but held in cell walls and intercellular spaces outside the protoplast, i.e. in the apoplast.

Cytoplasmic membranes contain between 30% and 50% water. Movement of water within, as well as across, these membranes occurs at all times. If the water potential within the protoplast is equal to that of the water in the apoplast, there will be no net movement, but merely diffusive exchange between the two regions. If, on the other hand, these water potentials differ, net flow in one direction will occur because of the universal tendency to establish equilibrium in water potentials. Whereas the water potential in the apoplast is mainly due to matric forces, and as a rule only to a slight extent due to osmotic forces, the water potentials of protoplasts have in addition to their matric and osmotic potentials a third component potential due to internal pressures; these pressure potentials often raise the water potential of the cell to its maximum, i.e. zero (p. 10). (N.B. Cells devoid of protoplast content can develop hydrostatic pressure potentials only, not turgor pressure potentials, cf. section 5.4.)

The composition of the water in the apoplast differs from tissue to tissue. In roots it is commonly a solute-depleted soil solution, owing to differential ion uptake across the plasmalemma into root cell protoplasts. In leaves it will be found to be almost devoid of solutes, as most of the ions from the xylem sap will have been translocated into parenchymatous cells associated with the vascular system in stems, petioles and leaves, or will have self-diffused back into the lower part of the plant. If this were not the case, mesophyll cell walls would become encrusted with salt deposits, with disastrous consequences. One can produce this situation by supplying a 0·75 M sodium chloride solution to the root system of a strongly transpiring plant. At this concentration salt will be delivered in the transpiration stream (p. 43) at vein endings and hydathodes in the leaf margin, where outward diffusion of vapour is most pronounced. Such strong salt solutions act as extra-protoplastic osmotica with exceptionally low osmotic potentials. As a result, water moves out of the protoplasts of mesophyll and epidermal cells; they collapse and leaf margins dry out becoming brittle; eventually deposits of salt crystals become visible (Expt. E5.3). This must not be confused with the salt deposits found on leaves of many halophytes and some species of *Atriplex*. This salt is secreted from the leaves by specialized salt glands and does not signify, as do the deposits mentioned above, that the plants are suffering severe water stress and osmotic injury. Indeed, in the case of plants which have salt glands, the deposits may be an advantage, in that they will reflect some of the light falling on the leaves and thereby lessen the radiation load.

Under natural conditions, however, the water potential in the apoplast is chiefly due to matric forces as has already been said. Whilst transpiration proceeds and water evaporates from walls into air spaces, the matric potentials decrease, sometimes to such an extent that it has been postulated that "incipient drying" may occur (pp. 43, 79) when transpiration rates greatly exceed rates of water flow to the leaves. Since the water system in the cell walls is said to be continuous from leaves to roots, the lowered water potentials due to matric forces will be transmitted through the xylem water system as a gradient in negative pressure potentials (p. 60).

5.2. Water movement across membranes

A truly semi-permeable membrane, such as a precipitated copper ferricyanide deposit in the walls of a clay pot (Pfeffer's classic membrane) is permeable to water only and impermeable to solutes. In plants, cell membranes are differentially permeable, i.e. their degree of permeability for water and different solutes is specific for each substance. When considering water movement across such membranes, which are permeable and permeated by water, it must be remembered that they nevertheless constitute a barrier and offer a resistance to free water movement. These resistances are reflected in the different permeability coefficients (see below) of membranes. The permeability of membranes to water changes with changes in temperature, the concentration of certain metabolites, of which carbon dioxide is noteworthy, and the presence of narcotic substances. These factors appear to affect the integrity of the membranes either directly or indirectly, by affecting the respiratory metabolism required for their structural maintenance.

In addition to diffusive flux of water across membranes there occurs osmotic flux whenever a membrane separates solutions of different water potentials. Without extending the theory of the derivation of the concept of "water potential" from the chemical potential of water, it can be seen from equation (1.2) on p. 10 and equation (A1) on p. 136 that the expressions for chemical potential of water contain terms for pressure and for concentration. Thus, differences both in pressure and in concentration represent driving forces for water movement across membranes. In the case of differentially permeable membranes, separate permeability coefficients for water and for solutes will influence their rates of movement. Moreover, solute movements have effects on the movement of water. When both pass together through membranes, the relationship and interactions between solute and water fluxes have to be considered in estimating the rate of the resultant osmotic flux. This is the function of the *reflection coefficient*. For a non-selective completely permeable membrane its value is zero, because there is no interaction between solute and water. If the membrane is permeable to water only and completely impermeable to solutes, the reflection coefficient will be 1·0, but for differentially permeable membranes it will have values between 0·0 and 1·0, depending on the specific permeabilities for the different solutes and for water.

As stated above, rates of osmotic flow depend on pressure and concentration differences. The magnitude of a hydrostatic pressure which would prevent osmotic flow can be calculated from osmotic concentrations of a solution. However, experimental measurements show that the pressure to be applied in order to stop osmotic flow is less than that expected from such calculations. This discrepancy is due to the interaction between the movement of solutes and of water referred to above, and characterized by the reflection coefficient which can thus be derived from the fraction:

$$\text{reflection coefficient} = \frac{\text{actual hydrostatic pressure preventing osmotic flow}}{\text{estimated pressure preventing osmotic flow}} < 1\cdot0$$

These considerations lead to an attempt to understand the nature of osmotic flux across membranes. The question is whether osmotic flow takes place by independent individual molecules, as in diffusion, or whether it involves an element of pressure flow. Experimental results show that water permeability coefficients of plant membranes, as determined by measurements of osmotic flow, are always greater (2–20 times) than determinations of these coefficients by diffusion rates of isotopically labelled water across the same membranes. This suggests that osmotic flow does involve bulk flow, caused by a pressure gradient across the membrane and traversing the pores of the membrane. It has been proposed that such a pressure gradient could result from the presence of solute at the solution side of the membrane. However, in the case of membranes permeable to some degree to the solute, its presence would not be restricted to the solution side but would extend in decreasing concentrations to the interior of the pores. Nevertheless, as the solution molecules (water and solute) undergo kinetic movement within the pore entrance on the solution side, water molecules from the more dilute solution just inside the pore will tend to diffuse out into the solution and thus cause the density of water to be lower at point A in figure 5.1 than at point B—hence the pressure profile within the pores of an osmotic membrane. The reduction in pressure has been calculated to be of the order of $(10^{-20} \times$ molar concentration of solute) bar. Such values, it is thought, would represent a sufficient driving force for osmotic bulk flow.

5.3. The osmotic component and the hydration of cytoplasm

Vacuolar sap contains organic and inorganic solutes, as well as colloidal substances, all of which confer upon the sap a solute or osmotic potential, often of considerable magnitude. These potentials represent the basic motive force for osmotic water movement but, as we shall see presently, by themselves they do not represent the water potential of the cell, which determines whether water will move into it or out of it.

The solute potential of the sap in most cells varies only between comparatively narrow limits, but in some cells, such as those of the mesophyll, phloem and stomatal apparatus, it can vary considerably. In this connection it is worth remembering that a given amount of water or solute movement will have a larger osmotic effect in cells with small vacuolar volumes than in those with large ones.

The vacuolar sap will tend to be in osmotic equilibrium with the

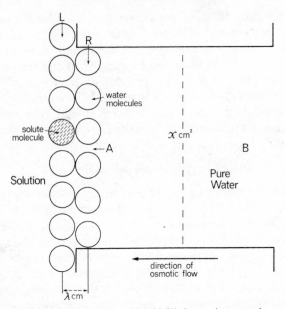

Figure 5.1. Idealized diagram of liquid-filled pore in a membrane. A sharp discontinuity between pure water filling the pore and solution is assumed at the solution end of the pore. Layer R, the pure water just inside the pore, has a lower density than the bulk pure water and therefore a lower pressure; and this is the source of the pressure which drives solvent through the pores and into the solution. (From: Dainty, J., (1965) *Osmotic Flow* in Symposium of the Society for Experimental Biology, XIX, 80–81).

cytoplasm that encloses it. The cytoplasmic lining of a vacuolated cell comprises the plasmalemma, the bulk of cytoplasm and the tonoplast, but these may be considered to function collectively as the membrane across which osmotic water inflow and outflow will occur, depending on whether the water held in the cell wall and intercellular spaces has a higher or lower water potential than the cell protoplast.

Besides being the motive force for osmotic water movement, the importance of the osmotic potential of vacuolar sap is that it determines the degree of hydration of the cytoplasm. For ecological studies and for considerations of the effects of the degree of hydration on metabolic processes, the relevant measurements are those of the osmotic potentials of sap (p. 119). This dependence of cytoplasmic hydration on the osmotic potential of the vacuolar sap is not altered by pressure potentials, which will be discussed below (p. 110).

As already indicated, the water potential of a cell is not determined by the osmotic potential of its sap alone, but also by the prevailing pressure potential within that cell. Such pressure potentials develop because cells, considered to be closed systems in the context of their water relations, must increase their pressure potentials when there is an osmotic flow of water into them, whereas any outflow must decrease their pressure potentials. Indeed, total cell water potentials are much more affected by changes in pressure potentials than by anything else.

5.4. The turgor component of the water potential of cells

The occurrence of (turgor) pressure potentials in vacuolated plant cells and tissues is of the utmost importance. Pressure potentials are the most variable component potentials affecting the water relations of cells, tissues and the whole plant. Pressure potentials in vacuolated plant cells may vary from zero to the same numerical value as the osmotic potentials of their vacuolar saps. The hydrostatic pressure potentials developing in non-living cells were dealt with in chapter 3; they are also very variable, but are of different origin and should not be confused with turgor pressure.

Pressure potentials referred to so far are positive potentials. Negative pressure potentials within cells (as distinct from those in the vascular system discussed in chapter 3) have been postulated, but their occurrence has not been demonstrated satisfactorily. They could arise from buckling of cell walls and the creation of a suction not unlike that existing in a compressed pipette bulb. Positive pressure potentials in cells, i.e. turgor pressures, result from osmotic water fluxes into cells. The consequent extension of the volume of the cell involves the stretching of its cellulose wall. This stretching requires the application of a force per unit area (pressure). This is the turgor pressure and it is opposed by an equal and opposite pressure, the so-called *wall pressure*. As the wall stretches, its resistance to further extension increases, so that the development of turgor pressure is thought to be not linearly related to volume changes (figure 5.2). Since turgor pressure develops as a consequence of osmotic water flux into the cell, the pressure under which this flux proceeds will be inversely related to the magnitude of the turgor pressure. This will be seen from figure 5.3 which summarizes these relationships; it can also be seen from the following numerical considerations.

For instance when the water potential of the vacuolar sap of a cell equals -5.0 bar (chiefly due to its osmotic potential) and the cell has a pressure potential of zero, the water potential of such a cell would be:

$$\psi_{cell} = \psi_p + \psi_{\pi sap} = 0 \cdot 0 + (-5 \cdot 0) = -5 \cdot 0 \, bar$$

If we assume the water in the cell wall and the intercellular spaces to have a water potential of $-1 \cdot 0$ bar, chiefly due to its matric potential, then water will flow into the cell with a pressure equal to the difference between the two potentials, i.e. $-5 \cdot 0$ and $-1 \cdot 0 = 4$ bar (differences between numbers have no sign). As the inflow continues, the sap will be diluted, raising its water potential slightly to $-4 \cdot 8$ bar, $-4 \cdot 6$ and $-4 \cdot 3$ bar respectively, while the outside source of water, if not voluminous, could develop a lower matric potential, so that its water potential will decrease to $-1 \cdot 1$, $-1 \cdot 2$ and $-1 \cdot 3$ bar respectively. As a result, the difference in water potentials between the two regions becomes smaller. At the same time a pressure potential will develop slowly at first, but at an increasing rate as the process continues.

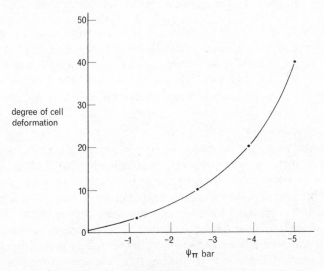

Figure 5.2. Deformations of a Nitella cell loaded with a constant weight under various osmotic pressures of the external medium. (From: Tazawa, M., and Kiyosawa, K., (1973), "Analysis of Transcellular Water Movement in Nitella", *Protoplasma*, **78**, 349–64.)

For instance when

$$\psi_p = +1 \, bar \quad and \quad \psi_\pi = -4 \cdot 8 \, bar, \quad \psi_{cell} = 1 \cdot 0 - 4 \cdot 8 = -3 \cdot 8 \, bar$$

and if $\psi_m = -1 \cdot 1$ bar, the force with which water will enter the cell will be equal to the difference between ψ_{cell} and ψ_m, i.e. $-3 \cdot 8 + 1 \cdot 1 = 2 \cdot 7$ bar.

At a later stage when

$$\psi_p = +2\,\text{bar} \quad \text{and} \quad \psi_\pi = -4\cdot6\,\text{bar}, \quad \psi_{\text{cell}} = 2\cdot0 - 4\cdot6 = -2\cdot6\,\text{bar}$$

and if $\psi_m = -1\cdot2\,\text{bar}$, the force with which water will enter the cell will be equal to the difference between ψ_{cell} and ψ_m, i.e. $-2\cdot6 + 1\cdot2 = 1\cdot4\,\text{bar}$. Finally, when

$$\psi_p = +3\,\text{bar} \quad \text{and} \quad \psi_\pi = -4\cdot3\,\text{bar}, \quad \psi_{\text{cell}} = 3\cdot0 - 4\cdot3 = -1\cdot3\,\text{bar}$$

and if $\psi_m = -1\cdot3\,\text{bar}$, the force with which water will enter the cell will be equal to the difference between ψ_{cell} and ψ_m, i.e. $-1\cdot3 + 1\cdot3 = \text{zero bar}$. Net flow of water into the cell will have ceased and the cell is as turgid as it can get because

$$\psi_{\text{cell}} = -1\cdot3\,\text{bar equals } \psi_m \text{ of the outside source.}$$

At all times the water potential of the cell will be given by the algebraic sum of the pressure potential and the osmotic potential of the vacuolar sap. In order to determine whether the cell will gain or lose water if in contact with an outside source of water, the water potential of that source must be known as well (cf. examples (a) and (b) below).

$$\psi_{\text{cell}} = \psi_p + \psi_{\pi\text{sap}}$$

The water potential of a cell is at its lowest when its pressure potential is zero, i.e. when the cell is incipiently flaccid. Plasmolysis (p. 120) can occur only in cells bathed in solutions with a lower osmotic potential than the osmotic potential of the cell sap. This situation rarely exists in nature. Leaf cells or root cells, for instance, do not plasmolyse, but become flaccid because, when the plant suffers water stress, the water in cell walls and intercellular spaces is held by strong matric forces which prevent its flowing into a space between cell wall and plasmalemma to permit the protoplast to plasmolyse, i.e. shrink away from the wall. Neither can air leak across cellulose walls to make such shrinkage possible. Cells which dry out are said to undergo cytorrhysis.

When the pressure potential reaches the same numerical value as the osmotic potential of the sap, the water potential of the cell is at its maximum of zero and no further net-inflow occurs. Positive water potentials of cells do not occur under natural conditions; they could be created only by the application of pressures on the cell. The importance

of the water potential of cells is that it indicates whether a cell will lose
or gain water when in contact with a source of water of which the water
potential is also known. Flow always occurs from the region with a
higher water potential to that with a lower. It must be emphasized that
this is not necessarily down an osmotic gradient, but down the gradient
in water potential. Examples (*a*) and (*b*) quoted below make this point.

(*a*) cell 1 with sap of lower osmotic   cell 2 with sap of higher osmotic
      potential but turgid   potential

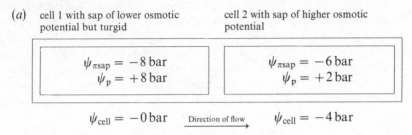

$$\psi_{cell} = -0\,bar \quad \text{Direction of flow} \quad \psi_{cell} = -4\,bar$$

Difference between cell 1 and 2 = 4 bar

Flow occurs from cell 1 to cell 2 with a pressure of 4 bar because
cell 1 has the higher water potential in spite of the lower osmotic
potential of its sap. The water in the apoplast which would be the
first reservoir from which water would flow into cell 2 has been
neglected. Equilibrium will be established when cell 2 reaches the
same ψ_{cell} as cell 1. We could estimate this to be at $\psi_{cell} = -2\cdot5$
when cell 1 has a ψ_π of $-8\cdot5$ bar and cell 2 one of $-5\cdot5$ bar on
account of dilution and concentration respectively, while ψ_p changed
to $6\cdot0$ and $3\cdot0$ bar.

(*b*) turgid cell with sap of lower osmotic   outside source with higher osmotic
      potential   potential

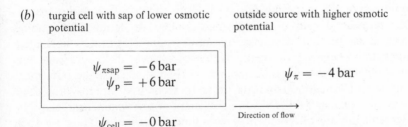

$$\psi_{cell} = -0\,bar$$

Difference between cell and outside source = 4 bar

Flow occurs from the cell to the outside source with a pressure
of 4 bar down the gradient in water potential in spite of the lower
osmotic potential of the cell sap.

Equilibrium will be established when the ψ_p in the cell reaches $+2\cdot5$ bar and $\psi_{\pi sap} = -6\cdot5$ bar owing to concentration. The outside solution is assumed to remain unchanged.

In both examples one cell was assumed fully turgid, but that need not be the starting point.

Before leaving the topic of turgor pressure it is worth while to draw attention to the fact that this internally generated pressure is exerted on the total cell content, i.e. plasmalemma, bulk cytoplasm, tonoplast and vacuolar sap. As a positive pressure potential it will therefore raise the component potentials (osmotic and matric) of all parts alike, so that *differences* in water potentials, e.g. between vacuolar sap and cytoplasm, will remain unaltered. The degree of hydration of the cytoplasm will continue to be a function of the difference between the water potentials of vacuole and cytoplasm. A dynamic equilibrium will be established when these potentials become equal. The lower the osmotic potential of the vacuolar sap, the less hydrated will the cytoplasm be at equilibrium. The development of turgor pressure does not alter this relation between the water potential of the sap and cytoplasmic hydration.

The most informative way to summarize cell water relations is Höfler's diagram (1922). Figure 5.3 is a version of this diagram which has been adjusted to our terminology.

5.5. Water relations of tissues

Tissues consist of interconnected cells functioning together. They can be simple and consist of the same type of cell throughout, or complex, i.e. composed of different kinds of cells. In both simple and complex tissues, cells may be devoid of cytoplasmic content or contain functioning protoplasts. In the latter case plasmodesmata constitute vital inter-connections between the cells. Water permeates all tissues and fills intercellular spaces. In some supporting tissues, such as the outermost cylinders of fibres in stems, the testa of seeds and strands of fibrous tissue associated with phloem, the amount of water held may be negligible but must become more voluminous at times to allow for such processes as, for instance, seed germination.

In most plant structures there are different kinds of tissues adjacent to one another. When in such structures water relations have equilibrated, all tissues will have the same water potentials, but osmotic potentials of vacuolar saps and turgor pressures prevailing within different tissues

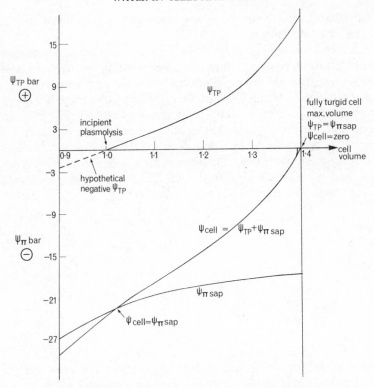

Figure 5.3. Höfler's diagram relating the water potential of a cell to its volume, to the cell turgor potential and to the cell sap osmotic potential. Modified to fit the terminology used in this book.

may differ greatly. The water in the apoplast will reach a uniform water potential before individual cells do. A section of stem may contain, amongst other tissues, parenchyma ground tissue, phloem and xylem. It may be assumed that the phloem sap is comparatively concentrated with an osmotic potential of, for example, -13 bar, while the parenchyma cells and xylem may contain saps with osmotic potentials of -6 bar and -3 bar respectively. Yet, if water is freely available, these tissues may all be fully turgid, with water potentials of zero, so that net water movement into these cells does not occur. However, under these conditions the sap in the apoplast remains mobile and, in response to a gradient in water potential between one section of stem and another, sap flux does occur. For instance, higher up the stem tissues may not be fully turgid, thus constituting the lower end of a gradient causing upward water movement.

Probably it is rare in actively growing plants for there to be complete equilibrium in water status, and some water movement is likely to occur at most times. In the example quoted above it was assumed that a gradient in water potential existed between lower and upper regions of a stem. It is important to realize that gradients are often to be found in one and the same tissue. The postulated water flux across a root cortex from root hair via endodermis to the stele is one such example which is familiar, but similar situations must be common in many tissues when an equilibrium is disturbed and flux into one cell of a tissue begins. This is shown below:

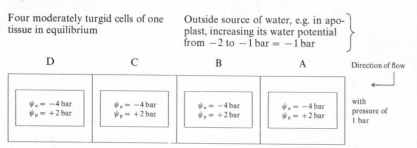

When the outside water potential changes, there will be water flux into cell A. This slightly lowers the osmotic potential of the sap in cell A and raises the cell's turgor pressure. When this happens, the water potential of cell A will be raised to above that of its neighbour, and in this way a gradient in water potentials will begin to develop from cell A to cells B, C and D.

In the above example the water potential of all four cells was -2 bar at equilibrium, and it was assumed that the water in cell walls and intercellular spaces was continuous with the outside solution without any additional matric potential. (This water in the apoplast would, of course, be available to each of cells A, B, C, D as soon as their water potentials fell.) Thus the flux need not be imagined as restricted to movement from cell to cell, but may occur also from the apoplast directly into cells.

Seeds. A special case in water relations is presented by seeds. Their tissue is dormant and its water content is unusually low. The water potential of a seed is almost entirely due to a very low matric potential, and the water that is present is tenaciously held by matric or adsorption forces of the colloidal system, whose continuous water phase is so scarce that probably all phases of the system will have changed from sol to gel. Membranes in a resting seed are thought to be partially disorganized, so that the seed is not an osmotic system. When a seed is exposed to suitable conditions of warmth and moisture, water uptake is due to the very steep gradient in water potential between the outside source and the very low matric potential of the seed. The process is called *imbibition* and, to begin with, water gained is immediately bound by adsorption forces to adsorption sites. This process is accompanied by release of considerable energy in the form of heat, as the adsorbed water molecules are restricted in their kinetic movement. Gradually, fewer and fewer adsorption sites remain vacant, and water is held by weaker electrostatic forces as the colloidal system becomes progressively more hydrated. Whilst this process takes place, cell membranes become reconstituted and begin to promote osmosis, so that water intake ceases to be due foremost to imbibition. It is the increase in water content of seeds which allows for these vital functions to begin which mark the change from the "resting stage" to that which we know as germination. Involvement of water in vital functions of metabolism are detailed in chapter 6.

Root Pressure. The phenomenon of root pressure, briefly referred to in chapter 3, is indeed the result of tissue water relations with which we deal here. Root pressure should not be looked upon as a cause of water movement, but rather as the result of water movement brought about by other forces.

Although other processes such as electro-osmosis may also be involved, it is certain that gradients in water potentials between soil solution and the root result in osmotic water flux into the root. This flux can be viewed, either as proceeding stepwise from cell to cell as in the example above, or as a movement from the soil solution with an osmotic potential of, for instance, -1 bar across a membrane system provided collectively by the cells of the cortex, endodermis and pericycle, into the xylem sap which must be assumed to have an osmotic potential lower than -1 bar. Such an osmotic potential may be due to organic or inorganic solutes contained in the sap. As a rule, the organic solute content of xylem sap is small, but the concentration of ions is usually sufficient to confer upon the sap an osmotic potential between -2 and -4 bar. It is thought that an active ion transport mechanism exists which delivers ions which were originally in the soil solution into the xylem sap.

As long as the gradient in water potential between soil solution and xylem sap exists, water will flow in, and this will gradually raise the columns of sap in the xylem so that a hydrostatic pressure potential develops. This pressure potential will counteract the osmotic forces responsible for water inflow, and will suppress it altogether once it equals the difference in water potential between soil and xylem. Thus, a manometer attached to a plant in this situation will measure a positive root pressure. If the plant begins to transpire reasonably vigorously, the positive pressure will gradually reduce to zero. The kind of negative pressure dealt with in chapter 3 will then develop (Expt. E5.1).

Guttation. When a plant is under root pressure, it can be assumed that all tissues have more than an adequate source of water in the apoplast, so that all cells can become fully turgid. Turgid cells may exert a certain pressure on the xylem elements, but being devoid of cytoplasmic content, xylem elements cannot develop any turgor pressure potentials. As a consequence of water flow into xylem elements, a positive hydrostatic pressure develops, the magnitude of which is basically a function of the height of the water columns in them. If, for instance, the difference in water potential between soil and xylem equals 1 bar then, theoretically, a column of 10 m height would exert the necessary hydrostatic pressure to prevent further inflow of water from the soil into the plant. However, as the major kinds of vegetation, e.g. grasses, herbs, shrubs and bushes are less than 6 m in height and are not hermetically sealed, such hydrostatic pressures cannot develop and water flows into the xylem system, even after all the tissues have become turgid. If there is no loss of water vapour from the aerial parts of the plant, liquid water will be seen to be extruded wherever possible. In most plants this is possible at the numerous vein-endings in the leaf margins (figure 5.4) or at special points known as *hydathodes*, water pores or water stomata. The latter are degenerated stomata which have lost or never developed the power to close their pore; an example can be found in the leaf of the strawberry. Often hydathodes are situated below a loosely packed irregular tissue known as *epithem*. Some hydathodes secrete salts in the guttation water, and on evaporation leave salt deposits on the leaf surface. Many of these salt deposits are weakly hygroscopic, and it is an open question whether they absorb water from the atmosphere or from the leaf—either acting as a means of removing water from the plant or

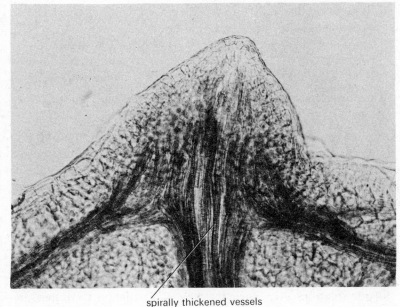

spirally thickened vessels

Figure 5.4. Photograph of a hydathode from the margin of a leaf of *Xanthium strumarium*.

providing a special boundary-layer effect, protecting the plant from further water vapour loss. Such hydathodes are found in the Saxifragaceae where they are called *lime glands*.

Relative Water Content and Water Saturation Deficit. The water status of leaf tissues is of special interest in many studies, and therefore special procedures are used to define the water status of leaves. One such procedure is the determination of the "Relative Water Content" by measuring the fresh weight of the experimental tissue before (FW) and after (MW) floating the tissue on water in a closed chamber for about 20 hours at 20 °C and at an illumination of about 2000 lux. After measuring the dry weight of the tissue the Relative Water Content is given by

$$R = \frac{FW - DW}{MW - DW}$$

Sometimes R is transformed into a percentage by multiplying by 100. A closely related concept is the "Water Saturation Deficit"; this is obtained by using the relative water content expressed as a percentage and subtracting from 100, i.e.

$$\text{Water Saturation Deficit} = 100 - \%R$$

5.6. The measurement of water potential and its component potentials in tissues

The Water Potential of Tissues. The *isopiestic* method of measuring water potentials described on p. 14 can be used with most tissues, and the *pressure bomb technique* described on p. 62 with material suitable for it. However, there is another simple and reliable method available which is suitable for many tissues.

Balancing Tissues in Graded Series of Osmotica. Materials suitable for this method are tissue strips or discs of roots, corms, tubers, fruit, many leaves and some stems. Strips are measured accurately with calipers, or discs are weighed, before immersion in different strengths of osmotic solutions. At regular intervals of 30 minutes these measurements are repeated, *until equilibrium values are obtained in all solutions.* Using sizeable tissue pieces, the length of times of immersion is not as critical as with the plasmolytic methods discussed below; errors introduced due to solute diffusion will be small.

Results of measurements are expressed as percentage changes at equilibrium, calculated on the first measurement, i.e. before immersion. The changes are plotted on the Y-axis against the osmotic potentials of the solutions on the X-axis. A fairly straight line is obtained which cuts the X-axis at the value of osmotic potential corresponding to the water potential of the tissue, i.e. the strength of solution which would have caused no change in the tissue.

With many tissues this method yields a rough value also for the osmotic potential of the vacuolar sap, where the plotted line shows a distinct kink or discontinuity in its slope below the X-axis. This is because once the tissue has lost its turgor, it will no longer shrink. In the case of weighing tissue discs, once the strong osmotica causing plasmolysis enter the space between cell walls and protoplast, the weight of the discs will no longer decrease, but begin to increase because the specific gravity of these osmotica exceeds that of the vacuolar sap— hence the discontinuity in the curves obtained. Example of results obtained with the leaf strips of *Vicia faba* are shown in figure 5.5.

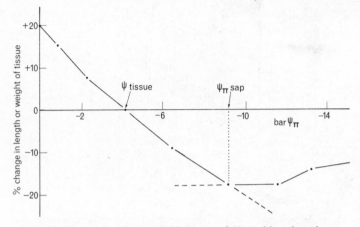

Figure 5.5. Results obtained with leaves of *Vicia faba* when determining leaf water potential by floating leaf strips in a series of osmotic solutions, and measuring changes in the dimension of the strips under a dissecting microscope fitted with measuring graticule. The discontinuity of the curve at -9.2 bar indicates the osmotic potential of leaf cell sap (for explanation see text). Other tissues can be used and changes in weight, instead of length, can be determined. (After: Kassam, A. H., *Hort. Res.*, **12**, 13–23, 1972.)

Shardakoff's Method. The water potential of plant material can be determined by placing samples of the tissue in a graded series of prepared solutions kept in small test-tubes. A second series of test-tubes is also required to provide reference solutions. These are coloured by the addition of a crystal of gentian violet. After the tissue has been kept in the test solutions and has established equilibrium by water exchange, a drop of the coloured reference solution is carefully placed below the surface of the test solution from a special pipette (figure 5.6). If the test solution has a lower water potential than the tissue, it will be diluted by water leaving the tissue, and the reference drop will be seen to fall from the pipette. In the opposite situation it will rise. That solution in which the reference drop diffuses symmetrically will have the same water potential as the tissue. If the tissue obscures the view, it can be removed before testing.

The Cryoscopic Method. The cryoscopic method has been used for the determination of the water potential of tissues. Some aspects of this method are discussed under osmotic potentials of sap (p. 120) for which it is more suited. However, if the fine measuring junction of a

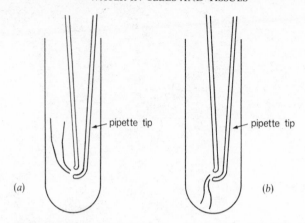

Figure 5.6. Diagrammatic indication of the experimental observations to be expected when using Shardakoff's method to determine leaf water potential. (a) ψ_{leaf} was lower than that of test solution; (b) ψ_{leaf} was higher than that of test solution.

thermocouple is implanted in a *small* volume of fresh tissue, and if a proper determination of the depression of freezing-point can be obtained, i.e. the sudden increase in temperature on freezing, then the value for the water potential calculated according to the expression shown on p. 120 will correspond to the water potential of the tissue. Often, however, it is not possible to supercool the tissue, because the numerous colloidal systems present have each their own freezing-points and interfere with each other in many ways. On no account should the same volume of tissue be used for replicate determinations, because it is certain that some substance will be precipitated or otherwise changed by the freezing treatment, so that the tissue properties will have been affected.

β-gauge Measurements. If a β-particle source, such as Tc-99, and a Geiger-Müller tube, sensitive to β-particles, are positioned so that they are on opposite sides of a leaf, and one facing the other, the leaf will absorb some of the β-particles; thus the number detected by the Geiger-Müller tube will be inversely related to leaf thickness. Since the main factor affecting leaf thickness (p. 31) in a mature leaf is the leaf water content, the readings from the Geiger-Müller tube can be calibrated in terms of leaf water content. The relationship between leaf water content and leaf water potential is non-linear. It varies considerably between species, and to some degree between different leaves of the same species. Nevertheless, the relationship can be determined by measuring the water potential with a thermocouple psychrometer (p. 15) and the water content by fresh weight and dry weight determinations. β-gauges can thus be used to determine leaf water potentials. The great advantage of this technique compared with many others is that measurements can be made frequently and non-destructively on one and the same leaf. Its disadvantages are that only the thickness of the intercostal areas of the leaf are measured, and that experimentally-determined correlations between thickness and water potential are not perfect. Estimated leaf water potentials will therefore be subject to inaccuracies.

Osmotic potentials

As was discussed on p. 105, the magnitude of the osmotic potential of cell sap is both of ecological and physiological interest, because this potential determines the degree of hydration of the cytoplasm.

Methods of measuring osmotic potentials of sap depend on the success with which samples of sap can be obtained. With large algal cells, sap can be withdrawn by means of micro-syringes or glass capillary tubes. Both techniques yield representative samples of uncontaminated vacuolar sap. In the case of cells of higher plants, the only practical method available is extraction under pressure. Such extracts suffer either from possible contamination by extra-cellular apoplastic sap and cytoplasm, or from being filtered through the cytoplasm, which may retain some components of the sap. Instead of pressure, centrifugal force can be used to extract sap from suitable cell material. This is achieved by placing the material on micro-mesh netting and centrifuging at such speeds as will break plasmalemma and cell wall; these will be retained on the mesh together with cytoplasm and organelles, while the sap is released. The purity of sap obtained in this manner is, however, not certain, e.g. it may be diluted by apoplastic water. Once sap has been obtained several methods are available.

The isopiestic method described on p. 14 is suitable for the determination of water potentials of any material and, since in the case of sap the only determinant of its water potential is the osmotic potential, this method is suitable.

Barger's method is another procedure based on vapour density relations. Droplets of sap are positioned in glass capillaries as shown in figure 5.7. On either side of the droplets are placed droplets of known osmotica. The assembly shown in figure 5.7 is observed under the

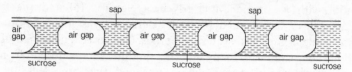

Figure 5.7. Arrangement of droplets of a known osmotic solution in a capillary tube, alternating with droplets of sap, when determining osmotic potentials of sap by Barger's method.

microscope, and measurements of the lengths of droplets are recorded. After 24 hours, during which the capillaries are kept at a uniform temperature, the measurements are repeated. If a droplet of sap has neither decreased nor increased in volume, it is deemed to have been positioned between osmotica of its own osmotic potential. The method is laborious but sound. It is based on water vapour diffusion, and condensation from solutions of higher saturation vapour densities to those of lower densities.

The cryoscopic method is based on the well-known depression of freezing-point, which is proportional to the osmotic potential of a solution. A drop of sap is supercooled to a known degree, and then seeded or otherwise made to freeze instantly; when freezing occurs, the latent heat of fusion (p. 4) is released, and the temperature rises dramatically to the freezing-point. The temperature measurements are best carried out with thermocouples, as these can be made small, and therefore need only small drops of sap. A correction for the degree of supercooling has to be applied to the calculation based on the relations:

$$\Delta\text{FP of a solution of } \psi = -22 \cdot 6 \text{ bar is } 1 \cdot 86\,^{\circ}\text{C}$$

$$\frac{22 \cdot 6}{1 \cdot 86} = 12 \cdot 12$$

$$\text{Therefore } \Delta\text{FP} \times 12 \cdot 12 = \text{osmotic potential of sap in bar}$$

Suitable instructions for this method can be found in textbooks of physical chemistry and also in *Water Deficits and Plant Growth*, Academic Press, 1968 (see also p. 24).

The Plasmolytic Method. The osmotic potential of vacuolar sap can be determined in intact cells. One method is known as *plasmolytic* and the other as *plasmometric*. Before going into details of these methods, a clear definition of *plasmolysis* must be given. A plasmolysed cell is characterized by the withdrawal of the protoplast from the cell wall, so that a space is created between the plasmalemma and the inside of the cell wall. Such a space can develop only by being filled with the plasmolysing solution permeating the cell wall. As was emphasized on p. 108, air cannot occupy such a space, because it cannot diffuse through the cell wall, and a vacuum does not develop in these situations. A plasmolysed cell can retain its vital functions for some time and, in experiments involving plasmolysis, cells should always be deplasmolysed by being placed in weaker solutions or water, in order to make certain that the determinations have been carried out with living cells.

50% plasmolysis is a method suitable for most tissues. Sections are placed in a graded series of osmotic solutions of known osmotic potentials. After about five minutes, the tissues are examined under the microscope, and the percentage of cells plasmolysed is determined in a number of microscope fields. When the results are plotted, the osmotic potential of a solution which would cause 50% plasmolysis can be obtained from the graph. This value is said to correspond to the mean osmotic potential of the sap in the tissue.

If plasmolysis is difficult to recognize, the tissue can first be placed in 1:10,000 Neutral Red dye, which will stain the vacuolar sap of living cells. This procedure is especially useful for the determination of *incipient plasmolysis* of single cells. In this method a section of tissue is placed successively in a series of solutions of known osmotic potential and observed. If a particular cell under investigation plasmolyses, it is deplasmolysed in a weaker solution until that solution is found which will just cause the beginnings of plasmolysis, i.e. incipient plasmolysis.

In the measurement of both 50% and incipient plasmolysis, solutions varying in osmotic potentials by as little as 0·1 bar can be used, so that the determinations can give "accurate" results. However, too high a claim for accuracy is not warranted because these methods have two weaknesses.

1. Some solutes will leave the cells by outward diffusion down their own concentration gradients or, if an inorganic osmoticum is used, some ions will diffuse into the cells. For these reasons plasmolytic work must be carried out speedily. To leave tissue for more than three minutes in the osmotica, irrespective of whether hyper- or hypo-tonic, will give vitiated results.

2. It is not certain that plasmolysis occurs when the osmotic potential of the vacuolar sap is barely exceeded by that of the outside solution.

It is possible that a rather stronger osmoticum is needed to cause plasmolytic withdrawal of the protoplast from the wall, because the plasmalemma may be held to the wall by plasmodesmata which might rupture only under excess force; hence the possibility exists that plasmolytic determinations of osmotic potentials of vacuolar saps yield exaggerated results.

The plasmometric method is a refined measuring technique requiring good microscopy and good micrometer measurements. It is difficult to apply. Cells are first plasmolysed in a fairly strong osmoticum. Volumes of the cell lumen and of the plasmolysed protoplast are next determined. For reliable volume determinations, cylindrical cells are best suited, and the form of plasmolysis must be convex. If these two conditions are given, the following relations permit the osmotic potential of the vacuolar sap at incipient plasmolysis to be calculated:

$$\psi_{\pi sap} = \frac{V_p}{V_c} \psi_{\pi osmoticum}$$

V_p = volume of plasmolysed protoplast, V_c = volume of cell lumen. Since $\frac{V_p}{V_c}$ is in fact a ratio, there is no need to express volumes in μm^3 but merely in (micrometer divisions)^3.

The Pressure Bomb Technique. Such measurements can be carried out by excising a shoot under water, and then placing it in a pressure bomb, with its cut end inserted into a water-filled flexible tube, the other end of which is suspended above a small container situated on a sensitive balance (figure 5.8). When a pressure is applied to the shoot, water will flow from the end of the tube, and the amount can be weighed. After a predetermined volume is expressed, the pressure is reduced to a value at which flow does not occur in either direction. This pressure is noted and is the balancing pressure. The pressure is then increased until the same volume of water is again expressed, and the balancing pressure recorded. If this procedure is repeated a number of times, and the total volume expressed at each pressure plotted against the reciprocal of that pressure, a graph will be obtained which is non-linear until the pressure is high enough for all cell turgor to be lost. At higher pressures, the graph will be linear, and extrapolation of this portion to zero volume expressed will indicate the initial osmotic potential of the cell sap.

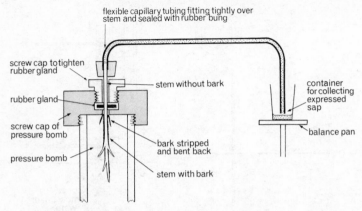

Figure 5.8. Arrangement of apparatus used to measure the osmotic potential of shoot tissues, using the pressure bomb.

(Turgor) Pressure Potentials. Until recently pressure potentials of tissues have been assessed only indirectly from the difference between their water potentials and the osmotic potentials of their vacuolar saps. Since

$$\psi_{cell} = \psi_{pressure} + \psi_{\pi sap}$$

the $\psi_{pressure}$ of a cell with a water potential of, for instance, -4 bar and an osmotic potential of its sap of -10 bar would be $+6$ bar.

Recently, pressure potentials have been measured directly by a turgor balance shown in figure 5.9, and even more directly by micro-injection needles which were placed into the vacuoles of individual cells and connected to pressure transducers and micro-syringes. The method depends on successfully implanting the micro-needles. This involves in the first place the puncturing of the cell, and then the reliable sealing-in of the needle by the plasmalemma, which is automatic and usually successful. As a rule the puncturing releases the *prevailing* pressure, and this measurement is therefore lost. But by applying pressure via the micro-syringe, it is easily possible to put the cell under pressure whilst measuring its changes in dimensions, and thus measure the pressure potentials of a cell which corresponds to measured changes of its volume (see Höfler's diagram, figure 5.3, p. 111).

Matric Potentials. Matric potentials of plant tissues are difficult to measure accurately, and most techniques are laborious. However, fairly reliable measures of pressure potentials of shoot and leaf tissue can be

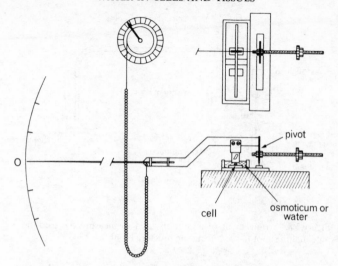

Figure 5.9. Side and plan view of a turgor balance. By adjustment of the chain weights, the indicator can be returned to zero and the (turgor) pressure measured. (From: Tazawa, M., and Kiyosawa, K., (1973), "Analysis of Transcellular Water Movement in Nitella," *Protoplasma*, **78**, 349–64.)

obtained with the pressure bomb (p. 63). Instead of inserting freshly cut leaf or shoot material into the pressure bomb, as is done when the water potentials are measured, for the measurement of the matric potential the tissue is first killed by freezing and then thawed before being placed in the apparatus. The pressure required to force water back along the shoot or leaf until it is just level with the cut end is equal to the matric potential of the tissue. Matric potentials measured in this way have been found to range from approximately zero bar in fully turgid tissues to about − 10 bar in slightly wilted ones.

Experiments

Experiment E5.1. *Demonstration of changes in pressure potentials in a simulated plant.*
The apparatus to be used is shown diagrammatically in figure E5.1; its assembly should be undertaken as follows. The thistle funnel and capillaries are filled by means of a syringe with a strong sugar solution and tested to see that the membrane does not leak. The glass tube with side-arm manometer is next filled with water and fitted to the thistle funnel, if necessary by keeping it under water. The porous pot must be boiled in water and filled with boiled water to the brim of the stem. This is then fitted on to the glass tube with manometer, if necessary keeping the joint under

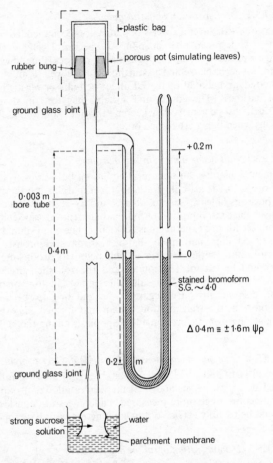

Figure E5.1. "Simulated plant" suitable for demonstrations of positive or negative pressure potentials resulting from "root pressure" or "high transpiration rates". $\Delta 0.4\,m \equiv 1.6\,m$ water pressure or $0.16\,bar$.

water, or turning the whole assembly upside down so as to avoid air bubbles entering the porous pot.

When the assembly is set up as shown in figure E5.1 with the plastic bag over the porous pot, a positive pressure will soon register in the manometer. After removal of the plastic bag, the porous pot will cease to sweat (guttate) and the manometer will soon begin to register reductions in pressure; gradually the positive pressure will change into a negative pressure. Gently playing a warm air fan on to the porous pot will speed up the procedure. These changes can be repeated several times.

Experiment E5.2. *Demonstration of guttation.*

Potted plants are used. Several species of plants are recommended so that guttation is seen to be a widespread phenomenon. All species of cereals, especially in the seedling stage are suitable; 30 cm high plants of cocklebur, dwarf bean, broad bean, tomato, tobacco, maize and nasturtium guttate well. Plants must be healthy and well watered, and should be placed under bell-jars or plastic hoods for about 12 hours of darkness. The covers and surface they rest on should make good contact; this can best be achieved by keeping this junction wet. Guttation droplets will be copious when lights are turned on.

Experiment E5.3. *Salt accumulation in leaf margins.*

This experiment illustrates an unnatural situation, but serves to demonstrate the point made on p. 102. It also illustrates that leaf margins are the most vigorously transpiring portions of leaves, because the boundary-layer resistance is much lower there than over the central portion of the leaf blade (p. 48), especially in still air.

A plant suitable for this experiment is cocklebur. The plant should be well watered and kept in good illumination, so that it transpires vigorously. By pouring a 0·75 M salt solution on the soil, leaves will begin to droop after a short time. The plant should be left for 15 minutes and the soil then drenched repeatedly with water to wash out most of the salt. After some time (one or two days) it will be seen that leaf margins have begun to die and dry up, because the salt has been swept there in the transpiration stream and formed an extra-cellular concentration of strong osmoticum in the leaf apoplast, killing the neighbouring cells.

Experiment E5.4. *Changes in leaf dimensions with leaf water content.*

Plants of broad bean and cocklebur are suitable. Leaf areas of fully turgid leaves are measured on graph paper, and the same leaves allowed to approach wilting, when their areas are once more measured. Differences are considerable. This experiment could be usefully carried out in conjunction with Experiment E2.1.

Experiment E5.5. *Demonstration of turgor movements.*

An excellent leaf for this experiment is a well-grown lupin leaf with pinnae of about 8–10 cm in length. Cut the full length of the petiole and allow the leaf to wilt, so that all the pinnae droop. Place the petiole in a large beaker full of water and cut off 3-cm lengths of the petiole in quick succession, until 15–20 cm have been removed. Leave the leaf in the beaker with water and observe. The full recovery of the leaf tissue and its fine form takes two to three minutes, and pinnae movements are easily seen as they occur.

Experiment E5.6. *Evidence of presence of moisture in the leaf apoplast.*

Cut strips of leaf tissue 0·5 × 4 cm from turgid leaves of broad bean, *Tradescantia* or *Commelina*. Place the strip across the index finger and cut with a sharp blade obliquely into the upper epidermis. Break the tissue at the cut and, gripping the broken tissue, gently pull off a strip of lower epidermis. Place the exposed mesophyll tissue on to a piece of filter paper and *very* gently press the tissue to the paper —the collected moisture will be seen on removal of the tissue. It is also most instructive to put the exposed mesophyll tissue under the microscope to study the disposition of the leaf air spaces. If upper epidermes can be stripped easily, the same experiment can be carried out with exposed palisade tissue.

CHAPTER SIX

THE ROLE OF WATER IN PLANTS

6.1. Involvement in structure

Water as Structural Component. Water is an all-pervading substance in much of the physical world and all of the plant world. Therefore we dealt with the state of water in the atmosphere, in the soil and in plants. We now conclude with a brief chapter listing some of the functions which water fulfils in plant structure, plant processes and plant metabolism.

In many plant organs water constitutes more than 90% of the fresh weight, and in few does water constitute less than 70%. The exceptions are seeds and other resting stages in life cycles of plants, in which the water content may be as low as 5%. Of these 5% most is adsorbed, and some is present as "bound water" not available for the functions of actively metabolizing plants. Indeed, the low water content of these structures has a good deal to do with their being dormant.

Water as Heat Sink. The low viscosity of water and its high degree of mobility, combined with its exceptionally high surface tension and strong cohesive forces, enable it to penetrate most capillary spaces, so that it establishes itself as a continuous medium throughout the skeleton of cellulose walls and totally permeates the plant body—indeed it forms a continuous phase in all plant colloidal systems. The presence of this all-pervading and relatively large volume of water in plant structures, especially shoots, acts as a heat sink and makes it possible for plants to withstand the continuous solar-radiation loads without attaining injurious temperatures. This is due to the high specific heat of water (p. 5).

The Role of Water in Plant Turgor. Although water forms a continuous phase within the plant, including its vital membrane system, it must be realized that different structural components possess different permeabilities for water and offer different resistances to the movement of water. This means that the plant water content is a dynamic osmotic system. Osmotic fluxes, combined with the virtual incompressibility of water, make it a medium well suited for the development of turgor pressure which is capable of giving a high degree of rigidity both to cell content and the enveloping cell wall. In herbaceous plants this represents a major skeletal force providing support for stems. In all plants turgor plays this role in leaves, flowers and fruits. This skeletal force is the more remarkable for not being static but "on the move", as we have discussed in the preceding chapters. That this is possible is due to the strong intermolecular forces which exist in water and which confer upon it a structural property, whilst at the same time it remains a mobile fluid.

6.2. Involvement in processes

Water as Translocation Medium. By translocation we understand the movement of substances from one locus in the plant to another. Such movement may be diffusional or aided by a metabolic change but, whichever it is, it depends on the presence of a suitable medium *in* which movement occurs. Movement through solids, when possible, is at all times slow; in the gaseous phase of matter movement is relatively fast but rather uncontrollable. There remains the liquid phase, provided in the plant by the continuous water system extending into sols and gels of colloidal systems.

Water as Transport Vehicle. In this context, transport of substances implies that they are moved *by* water and not merely in it. Water transport of substances in this sense does perhaps not play a very important role in plants, but it occurs. For instance, solutes having entered the plant from the soil solution may be aided in their further translocation (mainly diffusional or by active mechanisms) by being transported also at the rate of the transpiration stream and *by* it. (It must be emphasized that this is not a primary aspect of solute dispersion, which is a function of phloem metabolism.) Another example of transport in the sense we have defined is cytoplasmic transport of organelles such as chloroplasts, mitochondria, other inclusions and indeed solutes. All these are transported *by* the cytoplasm in its streaming movement which occurs to a greater or lesser degree in most cells. However, the rate of streaming is often not the same as the rate of movement of the organelles.

Water as Transpirant. Water itself is the only substance directly involved in transpiration. Its latent heat of vaporization (p. 4) makes the evaporation of water a cooling process of the greatest importance to plant life. Other cooling mechanisms, such as convection and re-radiation, are also of great importance; however, the leaf temperatures quoted on p. 29 show to what extent the cooling effect of transpiration prevents injurious temperatures being reached by leaves exposed to the rays of the sun.

The consequences of transpirational water-vapour loss have been dealt with at length in chapter 3. Here we merely refer to the transpiration stream as a realistic indication of the "turn-over" of the water content of plants. Thus it is due to transpiration that the plant water content is not a static but a dynamic system.

6.3. Involvement in metabolism

Water as Solvent. Perhaps the most fundamental role of water in plant metabolism is due to its being the best universal solvent known to us (p. 2). In addition to mineral salts dealt with on p. 2, most organic acids and their salts, which constitute much of the metabolic currency, notably in respiration, are readily soluble in water and some of them subject to weak ionization. Other organic substances such as sugars, which do not ionize, nevertheless dissolve in water; and those which do not form true solutions, such as many of the proteins, do form colloidal systems with water. These systems change from the sol to the gel state and vice versa with attendant energy exchanges. These changes in state contribute to causing cytoplasmic movements which are an important mixing mechanism for metabolites within cells. Thus many of the essentials for biochemical change are provided by the presence of water in cells.

The solubility of oxygen and carbon dioxide in water has been dealt with on p. 6 but must not be overlooked when considering the role of water in metabolism, since both photosynthesis and respiration depend on the solubilities of these gases in water.

Water as Reaction Medium. The very nature of many biochemical reactions depends on reactants being in the ionic form. Thus water in its role as universal solvent provides the medium in which these reactions can occur most readily. Reactants and products of reactions diffuse in solution in water. This allows for the Law of Mass Action to exercise its regulatory influence on metabolism via rates of translocation to and from reaction sites, and provides for metabolites to be moved away from reaction sites so that the reactions may continue at the required rates.

Water as a Source of Essential Ions. By its own degree of ionization (p. 2) water provides two of the most vital reactants, namely hydrogen and hydroxyl ions. To mention only one example involving protons, we refer to their part in the formation of reducing power (NADPH) during the photochemical phase of photosynthesis. As far as hydroxyl ions are concerned, not only do they provide the electrons required for the replenishment of depleted chlorophyll molecules, which are the source of excited electrons starting the photosynthetic conversion of radiant into chemical energy, but the hydroxyl ions are also the source of molecular hydrogen and of the oxygen which is evolved during photosynthesis. The oxygen represents the sole example in nature of the replacement of oxygen consumed from the atmosphere, i.e. water is the source of the vast amount of oxygen evolved by the mass of vegetation on land and in the oceans.

Water as Reactant. The undissociated water molecule itself is also an important biochemical reactant, especially in the many condensation and hydrolysis reactions of metabolism, such as the breakdown of polypeptides to amino acids or of polysaccharides to glucose and many other similar reactions. In this context we may mention the formation of metabolic water as one of the end products of respiration. In this process water molecules are formed *de novo*. In non-biological systems the formation of a water molecule is accompanied by the release of the energy of formation amounting to $16,000 \, J \, g^{-1}$. In biological systems this process is preceded by the electron transport chain leading to the terminal oxidase system. At several links of this chain, the energy of the transported electron is reduced and utilized in the formation of ATP, so that at the end, when metabolic water is finally formed, a much reduced energy gain occurs. Thus, the end point of processes driven by energy derived from sunlight, harnessed during photosynthesis, and partially made available during respiration, is associated with the formation of metabolic water.

6.4. "Water use efficiency" and plant productivity

Most plants, even those growing in "wet" environments, suffer some degree of water stress from time to time. This results in a fall in the rate of CO_2 fixation, either because the stomata close, or because the stress causes cell metabolism to change. Most plants can survive short dry periods without serious injury, but those that grow in arid or semi-arid environments often have to survive severe stress, and it is in these species that adaptations for drought survival can most easily be seen. Arid environments range from deserts to dry habitats in wet regions, e.g. those occupied by epiphytes in rain forests.

Plants inhabiting dry regions

Ephemerals: These may be described as drought-escaping plants. After rain has fallen, the seeds of ephemerals germinate, and the seedlings grow very rapidly. In many species flowers start to appear very soon after germination, and continue to be produced along the rapidly growing stems while the earlier flowers mature and produce seed.

Xerophytes: Plants which are drought-enduring, exhibiting structures that can reduce water loss and can also survive severe tissue desiccation, are known as *xerophytes*. Water loss may be reduced by the virtual elimination of cuticular transpiration, owing to the presence of hairs or of a thick cuticle or both; by the elimination of leaves or reduction in their size, e.g. in plants like broom; by the orientation of leaves in such a way that they intercept relatively little of the radiation during the hottest part of the day, e.g. in many Eucalyptus species; or by the ability of the leaves to change shape and roll up with the side of the leaf containing stomata on the inside of the cylinder that is so formed, e.g. in marram grass. All xerophytes can survive tissue desiccation to some extent, but one of the most striking examples of this is the creosote bush which occurs in some semi-arid regions of North America. The water content of viable leaves of this species can fall to below 50% of their dry weight, compared with normal water contents of about 200% for leaves of woody mesic species. The plants which can withstand most severe desiccation are the cryptogams, which can remain air-dry in drought periods but resume growth when water is re-supplied.

The fact that xerophytes have adaptations to reduce water loss does not mean that they necessarily have lower transpiration rates than mesophytes. In fact, when the water supply is plentiful, xerophytes often have a transpiration rate higher (on a leaf area basis) than that of mesophytes, and the former may dry the soil to such an extent that they ultimately cause themselves to become stressed. However, when water is in short supply, xerophytes can regulate loss much better than can mesophytes, by stomatal closure and the presence of a thick cuticle which may reduce water loss almost to zero.

Succulents: These form a category quite separate from the two above, in that they escape internal drought by storing large quantities of water in thin-walled cells, and have mechanisms to prevent water loss, but internal water deficits are harmful to them. When succulents are under water stress, their stomata close for all or most of the day, so that vapour loss is negligible because of the thick cuticle.

Their stomata are open at night when CO_2-fixation by PEP carboxylase proceeds. Any water which is lost during the night will come from water storage tissue, which will shrink in volume.

Photosynthetic Efficiency and Water Use. Even though succulents often keep their stomata closed during the day, they may nevertheless have an appreciable photosynthetic rate, as they open their stomata at night and fix CO_2 into malic acid. During the day malic acid is decarboxylated and the available CO_2 refixed by ribulose diphosphate in the Calvin cycle. This twofold process of incorporating CO_2 is known as Crassulacean Acid Metabolism (CAM). Plants which have this adaptation can also carry out photosynthesis in the normal way, and do so increasingly as the water supply improves and the difference in temperatures between day and night decreases. This means that the CAM pathway becomes increasingly important compared with the Calvin pathway, as the potential transpiration rate and drought increase.

The *transpiration ratio* (sometimes referred to as the *water-use efficiency*) of CAM plants is lower than that of any others, being 50–55 g water transpired per gram of dry matter produced, compared with 250–350 for C_4, and 450–950 for C_3 plants. However, when carrying out these comparisons, it must be remembered that the CAM pathway is the least efficient in terms of total dry-matter production.

Apart from the fact that during severe water stress xerophytes can greatly restrict their water loss compared to mesophytes under the same conditions, they do not have any advantage in water use economy as compared with the mesophytes.

Ephemerals are adapted to have a very high rate of productivity for a relatively short period, but this is at the expense of a large water usage. Many of them are little different in structure from mesophytes.

6.5. Water deficit and plants

When plants established themselves on relatively dry land, they evolved the ability to withstand some degree of water stress. Indeed, as was discussed in preceding chapters, especially chapter 3, movement of water into and within a land plant is a consequence of differences in water potentials. Often it is water stress that produces such differences in water potentials.

In land plants mild water stress is a common occurrence. More severe degrees of water stress, however, are often injurious and detrimental to healthy plant life. Thus, we mentioned in section 6.4 some adaptations which either overcome or evade these consequences in specially exposed habitats. Plants which lack such adaptations suffer as a consequence of severe water stress. It would be impossible to deal with this topic adequately in one small text on *Water and Plants* and therefore we include in our list of recommended reference books Kozlowski's *Water Deficits and Plant Growth*, and restrict ourselves to mentioning some of the important effects of water stress on plant metabolism which have not been touched upon in the preceding chapters.

Respiration. Rates of respiration are known both to decrease when water stress develops and to increase during water stress. Decreased respiration rates may be due to decreased cytoplasmic mobilities of substrates and end products, so that low concentrations of the latter and high ones of the former may result in slowing down the enzyme catalysed reactions of respiration.

The increased respiration rates during water stress seem to be related to increased rates of hydrolysis of starch to sugar due to greater amylase activity under stress; this would lower the water potential of cells and produce a greater tendency for water to move into the stressed parts. Other enzyme activities that have been observed to increase under water stress are those of ascorbic acid and catalase. In some species, such as cacti, there occurs an accumulation of polysaccharides and pentosans, but cacti are rather specialized plants.

Photosynthesis. Rates of photosynthesis are reduced by water stress for several reasons, one of which is related to stomatal functioning and leaf diffusion resistance to carbon dioxide which was mentioned in chapter 2. In addition to these effects there seem to be adverse effects on some of the steps of the photosynthetic process *per se*; for instance, the carboxylation efficiency of ribulose diphosphate carboxylase seems to decline under water stress. Coupled to this are decreased membrane permeabilities during water stress and lower rates of outward translocation of photosynthate in the phloem which may contribute to the general slowing down of photosynthesis.

Plant Growth and Composition. By its effect on the incidence of cell division water stress reduces growth rates. A decrease in water potential from zero to -1 bar may cause a reduction in the number of cell divisions per unit time by approximately 50%. It may be mentioned, however, that cell divisions continue to occur at a slow rate even in very severely stressed plants.

Growth depends also on cell enlargement, and this depends in turn on cell turgor pressure. Thus water stress directly affects rates of cell enlargement when cell turgor is reduced. An additional indirect effect on cell enlargement is the reduced rate of cellulose synthesis detrimental to wall formation. Another effect of water stress on plant composition is that the total carbohydrate content must be reduced when photosynthesis is reduced and respiration increased as mentioned above. Nitrate metabolism, too, is specifically affected by water stress, as proteolysis is speeded up in consequence of a shortage of oxygen when stomata remain closed during periods of stress. This is made more serious by a reduced rate of nitrogen incorporation into amino acids, resulting in an increased nitrate content when water stress prevails. This observation has been corroborated in that levels of amino acids in water-stressed plants remain depressed so that protein synthesis is sharply reduced.

APPENDIX

DERIVATION OF WATER POTENTIAL FROM CHEMICAL POTENTIAL

Recalling the Second Law of Thermodynamics (p. 7) ("Once a state of equilibrium has been attained in any process, the system will not *spontaneously* depart from that state of equilibrium"), it is clear that at equilibrium the entropy will be at its maximum value, and the internal energy of the system will remain constant. If we consider a system containing only one component, then the second law can be expressed as

$$dE = TdS - PdV \tag{A1}$$

where

$$dE = T \times dS - P \times dV$$

dE	T	dS	P	dV
change in internal energy (joules)	temperature (K)	change in entropy	pressure (bar)	change in volume (m³)

This means that the internal energy of the system will change by an infinitesimal amount if the system does an infinitesimal amount of expansion work. Equation A1 shows that only the internal energy, the entropy, and the volume need change if expansion work is done. This is not possible in practice, as a temperature or a pressure change or both must accompany a volume change, but the argument is simplified and the end result is not affected if the system is treated as a theoretical one in which neither of these parameters varies considerably.

If there is more than one component in the system, and there are n_i moles of component i and n_j moles of one or more other components, then equation A1 can be expressed as

$$dE = \left(\frac{\partial E}{\partial S}\right)_{V,n} dS + \left(\frac{\partial E}{\partial V}\right)_{S,n} dV + \sum_{i=1}^{N} \left(\frac{\partial E}{\partial n_i}\right)_{S,V,n_j} dn_i \tag{A2}$$

where N is the number of gram moles of N components making up the system. The subscript n means that the number of moles of each component is constant, and the subscript n_j means that the number of moles of all but component i are kept constant, so that the calculated value of dE in equation A2 is the value for the internal energy of component i in the system.

The chemical potential μ_i of component i is now defined as

$$\mu_i = \left(\frac{\partial E}{\partial n_i}\right)_{S,V,n_j} \tag{A3}$$

An analogous expression can be derived in terms of Gibbs free energy (G):

$$\mu_i = \left(\frac{\partial G}{\partial n_i}\right)_{T,P,n_j} = \bar{G}_i \tag{A4}$$

where $\bar{G}_i$ is the partial molar Gibbs free energy. The subscripts T, P, and n_j define their values as being constant (i.e. the entropy, the pressure, and the number of moles of all components are kept constant, except for component i (the component we are interested in). Under these conditions the chemical potential is equal to the rate of change of Gibbs free energy of a system with n_i moles of component i.

Chemical potential is an intensive property. Therefore, a difference between the chemical potentials of a substance in two phases (*phases* being understood thermodynamically, see p. 10) determines the direction of net diffusion of the substances between the phases.

At thermodynamic equilibrium, the vapour and liquid phases of a substance have identical chemical potentials. Therefore, the chemical potential of a liquid may be expressed in terms of its partial pressure in the vapour phase. Raoult's law states that the vapour pressure of a substance in solution is directly proportional to the mole fraction of that substance. In an ideal solution

$$p_i = X_i \, p_i^0 \tag{A5}$$

where p_i is the partial vapour pressure of the ith component, with a mole fraction X_i, and p_i^0 is the partial gas pressure of the pure form of the ith component at reference temperature and pressure. As the partial vapour pressure of water is related to relative humidity, this relationship between the vapour and liquid components is made use of in the measurement of water potentials—by measuring the equilibrium relative humidity, at a known temperature, of the air over a solution, a piece of plant material, or soil and calculating the unknown water potential (see p. 14).

Infinitesimal changes in the partial molar Gibbs free energy in a closed system depend on infinitesimal changes in temperature and pressure.

$$d\bar{G}_i = (\underset{\substack{\text{change in} \\ \text{partial} \\ \text{molar} \\ \text{Gibbs free} \\ \text{energy}}}{-\bar{S}_i} \times \underset{\substack{\text{partial} \\ \text{molar} \\ \text{entropy}}}{dT}) + (\underset{\substack{\text{partial} \\ \text{molar} \\ \text{volume}}}{V_i} \times \underset{\substack{\text{change} \\ \text{in} \\ \text{pressure}}}{dP}) \tag{A6}$$

At equilibrium at constant temperature

$$\left(\frac{\partial \bar{G}_i}{\partial P}\right) = V_i \tag{A7}$$

By integration

$$\bar{G}_i = \int_{P_2}^{P_1} \bar{V}_i(P,T)\,dP \qquad (dT = 0) \tag{A8}$$

Then if P_1 and P_2 are taken to be the partial vapour pressures of the ith component in the system (p_i), and of the ith component in the pure phase at reference temperature and pressure (p_i^0), respectively, then equation A8 becomes

$$\Delta \bar{G}_i = \int_{p_i^0}^{p_i} \bar{V}_i(P,T).dp_i \tag{A9}$$

From the ideal gas law

$$\bar{V}_i = \frac{RT}{p_i} \tag{A10}$$

Then from equations A9 and A10

$$\Delta \bar{G}_i = \int_{p_i^0}^{p_i} \frac{RT}{p_i}dp_i \tag{A11}$$

$$= RT \ln (p_i/p_i^0) \tag{A12}$$

When a substance is changed physically in some way from its pure state at reference temperature and pressure, then the change in partial Gibbs free energy is equal to the change in chemical potential:

$$\Delta \bar{G}_i = \mu_i - \mu_i^0 \tag{A13}$$

where μ_i^0 is the chemical potential of the pure form of the ith component at reference temperature and pressure. Then from equations A12 and A13

$$\mu_i - \mu_i^0 = RT \ln (p_i/p_i^0) \tag{A14}$$

From equations A5 and A14, the chemical potential of the ith component can be expressed in terms of its mole fraction:

$$\mu_i - \mu_i^0 = RT \ln X_i \tag{A15}$$

For non-ideal solutions (electrolyte or concentrated non-electrolyte) the equation is of the same form, but the activity coefficient ϕ, or the activity a, are substituted in equation A15.

$$\mu_i - \mu_i^0 = RT \ln \phi X_i \tag{A16}$$

$$\mu_i - \mu_i^0 = RT \ln a_i \tag{A17}$$

where

$$\mu_i - \mu_i^0 = \psi \tag{A18}$$

FURTHER READING

Esau, K., *Plant Anatomy*, Chapman & Hall Ltd., London, 1965.

Fogg, G. E. (ed.), Society for Experimental Biology, Symposium N°XIX, *The State and Movement of Water in Living Organisms*, Cambridge University Press, 1965.

Kozlowski, T. T. (ed.), *Water Deficits and Plant Growth*, Academic Press, London, 1968.

Kramer, P., *Plant and Soil Water Relationships*, McGraw-Hill, New York, 1969.

Meidner, H., and Mansfield, T. A., *Physiology of Stomata*, McGraw-Hill, London, 1968.

Monteith, J. L., *Principles of Environmental Physics*, Edward Arnold, London, 1973.

Penman, H. L., *Humidity*, The Institute of Physics, London, 1955.

Slatyer, R. O., *Plant-Water Relationships*, Academic Press, London, 1967.

Spanner, D. C., *Introduction to Thermodynamics*, Academic Press, London, 1964.

Wiebe, H. H., *et al.*, *Measurement of Plant and Soil Water Status*, Utah State University Press, 1971.

Index

148 INDEX

units of energy 6

vacuole, sap 108–112
vaporization 4–8
  heat of 95
vapour 6, 12
  deficit 39, 40
  lowering of vapour pressure 43, 44, 137
  phase 6
  pressure 8, 39, 43, 44, 136–139
variations, cyclic 64, 74–76
veins 50, 51
  endings 102, 114
  extensions 52
viscosity 5, 128
viscous drag 51
volume water flow 56, 57

wall pits 65
  pressure 106
water, adsorbed 3, 23, 87, 95, 113
  as heat sink 128
  as reactant 131
  as reaction media 130
  as solvent 2, 130
  as source of ions 131
  as translocation medium 129
  as transpirant 130
  as transport medium 129
  available 95
  balance in leaves 78
  bound 3, 113
  content of leaves 40, 41, 83, 129
    of soils 91, 96, 98
    relative 115
  deficit 134, 135

density 5, 58
flow, linear 12, 53–56, 62, 63
metabolic 131
movement, lateral 55, 80
  in membranes 102
  in roots 80
  in soils 89–94
potential 9–11, 42, 136–139
  gradient 112, 113
pure free 10
saturation deficit 79, 115
soil 81, 87–98
status 74–75, 95
stomata 114, 115
stress 28, 41, 42
table 93, 94
use efficiency 131
vapour 12, 30, 31
  deficit 39
  density 7, 8, 27, 32
  in soil 94
  pressure 136–139
wetting zone in soil 90
wilting percentage 96, 97
wind 26, 32, 34, 40, 95
wood 53, 83

xeromorphic leaves 44, 52, 132
xerophytes 49, 132
xylem 2, 50, 52, 58–67
  sap flow 56–58, 83
  shrinkage 60–62, 67

zone of root hairs 80, 112
  saturated of soil 90
  transmission 90
  wetting of soil 80, 112